CMMS

A Timesaving Implementation Process

Plant Engineering Series

Series Editor

Robert J. Latino

Vice President of Strategic Development
Reliability Center, Inc.
Hopewell, Virginia

FORTHCOMING TITLES

Results-Oriented Maintenance Management
Christer S. Idhammar

Total Equipment Asset Management
S. Bradley Peterson

PUBLISHED TITLES

Root Cause Analysis: Improving Performance for Bottom-Line Results, Second Edition
Robert J. Latino and Kenneth C. Latino

CMMS: A Timesaving Implementation Process
Daryl Mather

CMMS

A Timesaving Implementation Process

Daryl Mather

CRC PRESS

Boca Raton London New York Washington, D.C.

Library of Congress Cataloging-in-Publication Data

Mather, Daryl.
CMMS : a timesaving implementation process / Daryl Mather.
p. cm. -- (Plant engineering)
ISBN 0-8493-1359-7 (alk. paper)
1. Plant maintenance--Data processing. I. Title. II. Plant engineering series (Boca Raton, Fla.)

TS192 .M455 2002
658.2--dc21

2002073629

Direct all inquiries to CRC Press LLC, 2000 N.W. Corporate Blvd., Boca Raton, Florida 33431.

International Standard Book Number 0-8493-1359-7
Library of Congress Card Number 2002073629
Printed in the United States of America 1 2 3 4 5 6 7 8 9 0
Printed on acid-free paper

Dedication

To Maria Elena Salazar:
Your motivation and patience throughout
this process have made this possible.

Preface

This book is the result of years of implementations of EAM and CMMSs throughout various countries. During each of these implementations there was always intense debate and discussion. Many of the debates, discussions, and even the end results were repeated time and time again, often in the same company. In addition, many of the failures and common errors of large-scale maintenance system implementations were often repeated, such as a failure to predefine the systems and processes required for maintenance management, to adequately integrate systems training and process training, and to adequately define the roles and responsibilities of the members of the maintenance organization as they applied to the system's use and the application of maintenance processes.

The result of this was that a great deal of time and money was often spent going over the same subjects and themes again and again, as well as recurring errors of implementations. Many times, clients are driven by software vendors, instead of the other way around. The need to have a "template" for implementation became obvious. A template is needed to allow clients to define their processes and apply that definition to the selection and subsequent implementation and use of a system, rather than purchasing a system and spending a great deal of money to have consultants either modify the system or the client's practices to get to some level of compromise.

Each year, there are literally billions of dollars spent on the selection, implementation, and optimization of enterprise asset management/resource planning (EAM/ERP) and CMMSs. Many of these result in failure to deliver the required benefits and to meet the client's often poorly defined requirements. At the time this book is being written in 2002, there is a general feeling of disillusionment in industry regarding the implementation of large EAM and ERP systems. A major reason for this disillusionment is the perceived failure of the systems to achieve the results that are originally anticipated. The template approach was originally conceived in 1997 during an implementation in Australia. It has been refined through the years to a point where it is easily used to implement and optimize any CMMS implementation. The aims of this book are the following:

- Reduction of CMMS failures
- Reduction of consulting time for implementation of these systems

- Increased accuracy in the selection of CMMSs
- Increased control over CMMS implementations
- Increased ease of calculating possible returns on investment to easily set the limits of required investment

The system has now been used with numerous clients and with a high level of success every time. It is hoped that it will be useful to your company or career in the area of CMMS selection, implementation, and optimization, as well as to allow you to achieve strong results for your company and clients.

Daryl Mather
darylm@klaron.net
www.klaron.net

About the Author

Daryl Mather is the founder and principal consultant of Klaron in Mexico City. Originally from Australia, Daryl hails from that country's robust mining and oil and gas industries. Klaron is a maintenance engineering firm dedicated to enterprise asset management (EAM) system selections and implementations, as well as reliability centered maintenance and root cause analysis technologies.

During his career, Daryl has been dedicated to all facets of maintenance management and engineering, principally in the implementation and consulting areas associated with large-scale EAM system implementations. During this time, it became evident that there were a great many redundant tasks and mini-projects throughout each implementation. This redundancy led Daryl to the development of an implementation template, the subject of this book. He is also an RCM practitioner and Klaron is a representative of the Aladon network in Mexico.

Contents

chapter one

Introduction

Modern maintenance

During the past 20 years, the term "CMMS" (computerized maintenance management system) has become synonymous with productivity improvement and control of maintenance management processes. However, what do we really know regarding modern CMMSs? What are they today, and how have they evolved from the work-order systems that began to appear 20 to 30 years ago?

The tools and methodologies available today for managing maintenance and operations are truly astounding. When compared to the mentalities and methods of the 1960s and 1970s, it can easily be seen that we have rapidly come a very long way. It can also be seen that maintenance management is of vital strategic importance and is no longer the necessary evil that it was once considered.

Maintenance is now generally accepted as a means of gaining additional control of operational budgets and of greatly increasing a company's bottom line. When we consider that maintenance costs can make up 40 to 50% of operational budgets in capital-intensive industries, the effect of a reduction in maintenance costs is both obvious and impressive. In addition, maintenance is generally the largest controllable operating cost of a capital-intensive industry. As such, the ramifications of slight improvements will greatly benefit the organization in the long term.

There has been a boom in the creation and supply of technologies for refining and optimizing the way the maintenance function can be delivered today. These are, of course, varied and diverse changes but generally they can be described within three broad categories: advanced methodologies, predictive technologies, and advanced management techniques.

Advanced methodologies

The development and acceptance of RCM (Reliability Control Maintenance) as a standard for maintenance strategy development has greatly increased the ability to control maintenance costs. Streamlined and varied approaches of the same style, such as preventative maintenance optimization, have also greatly enhanced our abilities in this respect.

The adoption and recognition of operational and organizational strategies, such as TPM (Total Productive Maintenance) and its predecessor TQM (Total Quality Management), have changed the way in which we view our operations. The Six Sigma methods and the evolution of root cause failure analysis as practiced and developed by many companies throughout the world are beginning to find their way into the maintenance environment.

Predictive technologies

Aligned with the methodologies previously discussed, there has also been a boom in the way we measure the performance of our equipment. This has been both in the testing that we do and the way in which we are able to do it. Some examples may include:

- Advances in predictive analysis such as thermography, oil analysis techniques, ultrasonics, vibration analysis, and a great many other areas
- The advent of online monitoring and alarm raising via a multitude of plant and equipment monitoring systems

Advanced management techniques

Here we begin to get into the topic of CMMSs. The modern CMMS is a very advanced tool for managing maintenance. Years of analysis and development by various organizations have led us to the point where we are able to address basically every requirement of maintenance management in a manner that is efficient and effective.

A modern CMMS is capable of handling all of the various processes and procedures of maintenance management, assisting the organization in making the operations more efficient, and analyzing equipment to further optimize performance in that area.

We also find in this area a proliferation of niche maintenance managment and analysis tools. These allow us to better apply some of the advanced maintenance methods and technologies in a practical and efficient manner. Some examples here may include:

- RCM and preventative maintenance optimization software tools
- RCFA (Root Cause Failure Analysis) software tools
- Technical change-management tools
- Inventory optimization and vendor relationship managers
- Logistics management (route calculations and other linear optimization methods)

Yet with all of these advances available to us, we still find ourselves in a situation similar to our counterparts in the 1960s and 1970s. We are trapped in a reactive maintenance environment with daily battles to keep in the operational mode. Why is this so? We have the technology available to us. We are also generally much better educated than our predecessors were. Our engineers have the very latest in technology available to them and have been taught the very latest in methodologies. Our managers are trained in advanced business management techniques, which are generally more advanced than they were in the past. As noted previously, often we can have the support of our superiors at the corporate level. So why are we still finding ourselves stuck in a reactive mode of maintenance management?

The answer is simple: The maintenance of business processes, and the management of them, have not kept pace with the rapid advances in technology.

This is not to say that the CMMS on the market today does not provide the ability to manage the advanced business processes we require. They do. However, we are not sufficiently prepared to fully utilize them. The functionality of these systems has outstripped our ability to manage our processes to suit.

So how do we measure maintenance performance, and where we are in terms of development of our maintenance capabilities?

Maintenance development

Maintenance is a process involving not merely a department with a mandate to "fix it when it breaks" but the entire organization as well. The process of maintenance primarily involves an evolution through three stages of development, each with its own identifying characteristics. The following descriptions are an overview only and will be further defined in the course of this book.

Reactive maintenance state

- Many uncontrolled maintenance stores, due to the inability to accurately predict and control the maintenance requirements.
- Large inventory holdings (related to the previous point).
- Reactive firefighting state of operations, which can be the hardest of things to change within a maintenance operation. Quite often the adrenalin of racing off to fix a problem, doing so rapidly, and being congratulated for doing so by the department head is hard to get employees away from. The "white knight" syndrome can be difficult to penetrate. There is resistance as employees who are very good at repairing problems rapidly may see themselves becoming superfluous. A reactive maintenance environment can also lead to demoralization of the workforce and a feeling of helplessness.

- Low MTTR (mean time to repair). This is a good thing; however, the fact that maintenance teams repair similar problems on the machines regularly sometimes means that they are well practiced. It is not a feature of good planning. One of the challenges when advancing through maintenance evolution is to maintain MTTR at low levels by replacing individual knowledge with overall planning.
- Low MTBF (mean time between failures). This often can be masked by high availability measurements. Further explanation of this is detailed later in this book.
- Excessive maintenance costs. In this respect particularly, indirect maintenance costs are extreme and often unmeasured. Costs include lack of availability, nuisance trips or faults, and the high inventory to manage all of this.

Planned maintenance state

- Capacity resource scheduling, better resource utilization, and fewer resources required
- Reduced inventory holdings
- Reduced number of uncontrolled maintenance stores
- Better team morale
- Better planned and scheduled work ratios, more planned and scheduled work
- Defined business processes and measures
- Understanding direct and indirect costs incurred by maintenance
- Increasing level of confidence from the operations departments and within the maintenance department
- Importance given to continuous improvement projects and efforts

Proactive maintenance state

- Optimized resource levels
- Optimized inventory holdings
- Majority of maintenance tasks are performance or condition based, little or no time-based maintenance
- Strong emphasis on reliability engineering and other continuous improvement processes
- Niche software programs added to enable even further fine tuning of the maintenance delivery effort
- Full operations involvement in the maintenance process (TPM)
- Low reliance on personnel, strong reliance on positions, strong documentation, and skills assessment training regimes (focused on corporate requirements)

However, prospective buyers of CMMS are becoming a lot smarter. They specify requirements exactly, they state their processes with clarity, and they generally supply information as to the benefits they want to derive. So why do they not get the results they expect?

Principally, it is due to poor or unprepared implementation and selection processes, and generally poor use of the system. However, a common explanation in the implementation business is that "the results of the implementation depend greatly on the consultant doing the work." The success of an implementation depends on the individual consultant, not the consulting company nor the software product, although obviously both of these are important. As a result, we see that the data contained in the CMMS or maintenance management module of an enterprise-level system can be very poor in content and quality. This is regularly supported by audits and surveys of the data in the CMMS.

Obviously, there are various reasons why this can occur, ranging from poor implementation or purchase to poor corporate support and follow-through. It is essential to note here, as we will discuss later in the book, that both the sponsorship and leadership of any implementation must be strong and based in belief and understanding of the benefits to be derived and how they will be achieved.

So we see that those consultants with a great deal of experience, which unfortunately must involve failures as well as successes, are the ones that are best able to get results for clients in this area (and most other areas of consulting, I would add). Some software and consultancy organizations have been able to standardize delivery of their products. In these cases, there is some measure of consistency; however, there is still the factor of comprehension and the ability to communicate this to the prospective client.

The aim of this book is to share some of the learned experience of a consultant in this field to enable you to make decisions on the use of the system with the global view of its benefits in mind. In short, this book will give you an implementation template, the ability to apply learned principles to the implementation of a CMMS in any operation.

There is no end of information on the CMMS selection process, and we will cover this later in this book. However, very few approaches actually speak about the preselection process and what needs to be considered prior to beginning the selection of a new CMMS.

The purpose of the implementation template is to empower the prospective CMMS buyer. The purchase and implementation of a CMMS is an opportunity to revise the maintenance business processes that are in place. What is required is an idea of what business processes need to be revised and the guidelines to do so. With the business processes redefined and optimized, you are then in a position to use this information as the selection criteria for your CMMS search. In addition, with these definitions in place you are in a position to drive the implementation process, rather than having the implementation process drive you. This is one of the key elements in the battle to prevent CMMS failure: knowing exactly what is required, and exactly how you will apply that to your company.

Baseline functionalities

Look at the requirements being placed on CMMS producers in terms of a baseline of functionality that is now expected from them. The days of a straight work-order system are long gone and have been replaced with comprehensive functionality involving various areas of the operational spheres. These generally need to cross the following areas:

- Maintenance control
- Inventory control and purchasing
- Human resources programming
- Financial reporting

It is within these functions that we begin to see the differences between CMMS/MRO (maintenance repair and overhaul) systems and their larger enterprise-level counterparts. A CMMS as such is generally focused only on the MRO functions of the operations, and the functionality within other "modules" is restricted to supporting these requirements. In the larger enterprise models, it generally goes much further and extends into the realm of workforce planning and development, financial analysis, and payroll, as well as advanced methods of inventory control and demand management. Recent products in the enterprise spheres are also becoming increasingly focused on E-collaboration issues. This is an exciting new area that will inevitably lead to even greater levels of productivity.

As a baseline, the following functions are not only good to have but a basic requirement of any system — the basic suite of functions that are required. Obviously, the enterprise-level systems as detailed previously have a requirement for much greater functionality. Suppliers that are not providing these as basic functions are "temporary businesses," and will not be able to market their products much longer within the current, changing environment.

Maintenance control

- Equipment details and statistics
- Abilities for both planning and scheduling
- Abilities for autoscheduling of recurring tasks
- The ability of integration with condition-based maintenance and its management
- Control of valuable or operationally critical components of equipment

Inventory control and purchasing

- Ability to create requisitions from the maintenance store
- Ability to forecast demand and detail this in costs
- Ability to control purchasing requirements in a demand-management capacity

Human resources programming

- Ability to capacity program a company's resources into the future (this obviously involves a full range of human resources data including details, rosters, and abilities and skills of each resource)

Financial reporting

- Ability to report via assets, cost centers or accounting codes, and accounting periods
- Package must allow configuration to align with the business processes and levels of authorization

It is not the easiest of tasks to be a provider in today's market. Currently, there are a great number of CMMS-style systems in the marketplace, each with a different degree of compliance with the basic functions of financial reporting. As clients of these systems become more aware not only of the importance of the investments they are making but also of the levels of functionality available to them, there is the inevitable adjustment of the number of providers and suppliers of such products. Due to the number of options available, a market that originally was very much focused in favor of suppliers and their products is now becoming very much driven by clients and their requirements.

Different systems

We have spoken briefly of the different systems available in the marketplace today and what defines them. There are a few highly reputable organizations dedicated to defining the different levels of products and determining how they are operating within each of these market and functionality areas. This level of detail will not be discussed here.

The systems now used to manage corporations at the enterprise level are vastly different in terms of their origins and current-day functionality. However, as they converge to achieve the principles of ERP II (enterprise resource planning), these differences will become less important and they will be identified more by market sectors than functionalities. In the future, we will have only enterprise management systems (EMS) and not the array of systems present in the market today.

In discussing the evolution of enterprise-level systems, it is important to note that they originate from different industries and were originally created for different requirements. They are now crossing these boundaries, although in some cases not very well (a factor which will determine the systems that will dominate the marketplace of the future).

There have been two distinct paths leading to the development of enterprise management systems; one is based on the requirement to manage physical assets, and the other is based on the requirement to manage production and its demands. These requirements can be stated as coming from two basic industry groupings: capital-intensive industry (enterprise asset management, EAM) and manufacturing and production-based industry (enterprise resource planning, ERP). At this point in time, with functionalities being what they are, it is very important from a selection point of view that you are able to categorize your operations within one of these two spheres. As such, you will have a strong idea of what your base requirements are in terms of the overall concepts.

What are the differences in these two industries that cause them to spawn unique operations methods and management systems that grow into powerful forces shaping the business community?

Capital-intensive industries

Capital-intensive industries require a high level of availability and reliability from their fixed assets. These industries are not governed by their production plans; in reality, their production plans govern the assets.

A good example of this style of industry is the mining sector. A mining or mineral processing operation will generally produce to capacity. The plant or physical asset base is defined according to the capacity it is required to produce. Client requirements are generally somewhat fixed, determined by ore body and quality of the product being mined, supply, and market. As such, the requirement on resources is that they will run to capacity for as long as they are able. Due to the large capital investment required to increase the capacity of such operations, there is a need to reach and occasionally exceed design capacity with the assets they currently have. The most profitable form in which to manage such an asset body is to operate it at capacity continually.

Another case of capital-intensive industry is the transport industry. Business requirements determine the fleet size that a transportation enterprise is able to operate profitably, and overall profitability depends on the fleet being available and reliable enough to operate at capacity on a continual basis.

Other capital-intensive industries include:

- Water and electricity utilities
- Railroads
- Military establishments
- Oil and gas installations
- Steel providers (to a lesser degree)

These industries rely greatly on the abilities of their assets to fulfill their requirements on a continual, on-demand basis.

Among the defining differences between these two industry groups, apart from the focus on maintenance management, are the distinct ways of purchasing and inventory management. For example, the ERP-based industries are very much advised to utilize a form of just-in-time (JIT) inventory management. However, the EAM industries are better suited to a very defined and specific form of just-in-case inventory management. There are a myriad of reasons for this, which will be detailed later; however, one of the stronger reasons is the lack of supplier diversity within the providers of materials to EAM-style industries.

An example of these types of providers is a heavy-equipment manufacturer or a manufacturer of specialized valves. There is limited demand for these types of products, few suppliers, and the market is not overly competitive; therefore, gaining compliance for JIT or other advanced inventory management techniques is not always easy or possible.

Manufacturing and production-based industries

This is a very different industry grouping with very definite requirements. The requirements of this industry group have driven the development of the materials resource planning (MRP) methodology and its latest incarnation, ERP. Although ERP is immediately recognized as an enterprise-level management system, it must be remembered that it is also primarily a production-planning methodology and strategy — the most effective of its kind in the world today.

As stated, one of the key areas where these systems differ is in their approach to inventory management. An ERP-style industry must plan purchases and inventory management around its production plan. The production plan is generally variable and can change rapidly with changes in market conditions or client requirements. As such, the inventory of an ERP-style company is attuned to the requirements of the business plan; in other words, the company purchases as and when required in order to produce a finished product. In EAM-style industries, requisitioning and inventory control are effected to satisfy the requirements of maintenance management and to

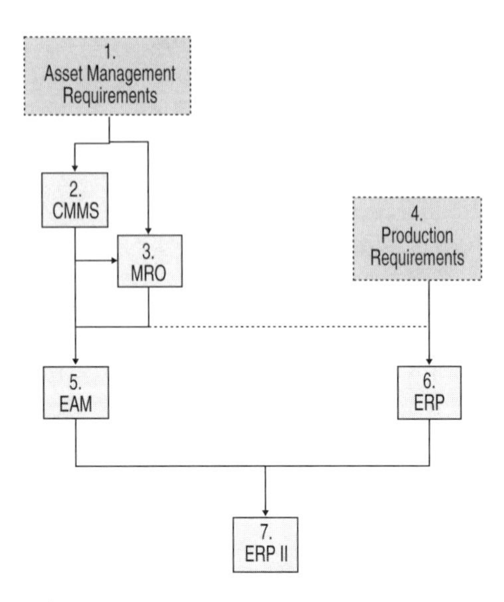

Figure 1.1 Evolution of management systems.

maintain equipment at financially acceptable levels of reliability and availability.

It is important to note that maintenance management does not form a part of the ERP or MRP methodology. As such, the systems built to manage these industries do not have maintenance management as one of their core functions; rather, they have the production plan and its management as core functions. However, that is not to say that these industries do not have strong maintenance management modules. They do, and some have even adopted the inventory management characteristics of the EAM-style industries and solutions providers to a limited degree. This factor alone, as well as some pretty impressive marketing, has allowed this style of system to coexist in the EAM sphere with relative success. But it is not, and never has been, at the center of their focus.

The range of industries in this area is vast and encompasses the majority of today's large industries, particularly within the U.S. and Europe.

As seen in Figure 1.1, these systems have followed different evolutionary paths to where they are today. The following definitions are provided in order to get a feel for how they evolved and what their focuses are in today's marketplace. The intention here is not to define each category exactly, nor to define which of today's systems belong in any category.

Box 1: Asset management requirements

Early in the development of enterprise management systems, there was a recognized need to manage the requirements of physical assets, and, although growing out of the capital-intensive sector, it was a need recognized by many maintenance practitioners without regard to their particular industry or its requirements. Maintenance was beginning to change, and the need to have better-organized structures for ever-growing requirements was obvious. In addition, the popularity of personal computers was opening the door to development in this area.

Boxes 2 and 3: CMMS and MRO

CMMS was the first of the systems to appear for managing the requirements of maintenance. In the beginning, these were very basic work-order management systems, sometimes linked to inventory functionality. They were designed to replace or mimic the work-order job card systems that were in use at the time (and still are in some installations). Basically, they were designed to bring into place an electronic means of controlling, planning, and analyzing the performance of the maintenance function or department. From this came the MRO-style systems. These were based in the CMMS style of thinking but focused more on the requirements for advanced planning and scheduling of tasks, advanced inventory management and budgeting, workforce requirements and planning, and the ability to manage other areas such as shutdown and modification management. These were also starting to become very focused on the ability of the system to not only report on costs but to drive cost reduction through greater control and analysis methods.

Today, both of these products are still in existence and continue to provide very strong service to maintenance departments and organizations that require a very specific maintenance focus rather than focusing on the overall enterprise management aspects. A market sector is definitely the small- to medium-range areas. Although both of these terms are still widely in use, the more common of the two is CMMS.

It is also worth noting that many enterprises look at integrating CMMS with an enterprise-level software system in order to develop a best-of-breed system, or something that they feel is ideally suited to their needs. This is a trend that is set to increase as systems improve the functional depth of their products along specific market positioning lines.

Best-of-breed systems are also put together out of a supplier's inability, or perceived inability, to meet the requirements of an organization. An ERP is a good example. Often, there are best-of-breed systems put together using an ERP system and a CMMS with good functionality. Although ERP systems have grown out of the production requirement of manufacturing organizations, one of their strong areas has been the ability to report on and control the financial aspects of an organization.

Box 4: Production requirements

At about the same time, there was the recognition that certain industries had very specific requirements in terms of production planning and control of their resources. The profitability offered to such industries by being able to better plan and control the inflows of these resources according to plan was starting to become both accepted and applied. At this time, there was a wide range of systems focused on various forms of applications, and it was from these that the core of the ERP-style systems originated.

Box 5: EAM (enterprise asset management)

EAM systems are direct descendants of the CMMS- and MRO-style systems. They remained true to the original focus on maintenance and maintenance-driven inventory control, and began to add the functionality now taken for granted within systems at this level. These were basically functions within the areas of human-resources management and planning, payroll capabilities, and extensive financial management capabilities. Over time, their maintenance abilities also became far more advanced than their early predecessors and the ability for capital-intensive industries to gain large benefits from these systems was greatly expanded.

Box 6: ERP (enterprise resource planning)

ERP systems also began to emerge around the same time. Instead of the maintenance and asset management base, they had originated from the production planning base. However, in line with the advances in EAM systems, there was the recognized need to cater to the broad requirements of enterprises in their industry groupings. However, the needs of their industry bases caused even greater early advancement than those of the EAM industry groupings.

As such, there was also strong development in the additional areas of human-resources planning and management, advanced administrative functions, commercial applications, and even more advanced financial management. The ERP is the dominant system in the world today, so much so that the functionality differences between it and the EAM-focused solutions are not widely known and not considered in selection processes.

Box 7: ERP II

ERP II is a new category which has yet to be fully realized by any of the major providers in the world as both the industry-style solutions are converging to become the enterprise management systems mentioned earlier. The sector also relates to extensive E-collaboration and interconnectivity with other applications. As such, there has become a requirement for truly open architecture at a system level and a focus on being able to interact with Internet marketplaces and other requirements.

chapter two

Defining the implementation requirements and scope

Returns on investment

Now that you know what it is that we are talking about, why do you need it? What are the potential returns on your investment, and specifically what is it that your business requires? This stage in the process is a good time to define exactly what it is that the business does, and how to define processes within the business. From this point forward, we will focus on CMMSs and the maintenance implementation of enterprise-level systems.

It is recognized that in any large system the overall process requires a focus on much more than the maintenance and operations content. However, it is within this area that the greatest time and cost savings, including indirect costs, can be achieved.

As always, there are a range of things that must be considered in order to move to the selection of your system. This chapter is intended as a guide to defining these and to defining the price range for your system. Obviously, there are other constraints such as the operational state of the enterprise and the level of urgency that is required.

First and foremost, why do you need this new system? And when you get it, what do you expect to get in return? This can be a very difficult area to prove at times. Organizations that are reliant on physical assets and operate without a CMMS are often convinced that they are doing a good or average job and cannot see the value in adding new software and responsibilities to their organizational structures. But what are these beliefs based on? Generally nothing more than anecdotal information and the resistance to change, I have found. There is rarely any comparison in terms of benchmarking to other companies within the industry, and there is rarely any evidence of continuous improvement. In fact, there can often be a resistance to the fact that they are stuck in a reactive maintenance mode and are not progressing at all.

Before we continue, it is essential to note that there is no system in the world that will deliver any return on investment (ROI) without the involvement of management at the corporate level. A strong and focused leadership effort is a key ingredient of any CMMS implementation and success.

A general assumption can be made here: CMMS will improve the maintenance efforts of organizations in the reactive and even planned states of maintenance. If there is no system in use today, there is a strong possibility that this operation is firmly in the reactive maintenance stage, regardless of any card system in place. The ability to analyze, plan, and schedule is compromised without an electronic medium.

Another fact worth mentioning is that a planned and scheduled maintenance task is at least 50% less costly in terms of dollars and time. So one of the benefits that can be realized immediately is the reduction of maintenance costs associated with unplanned and unscheduled work. This is also an area where you need to be absolutely frank regarding your performance. What was the percentage of work done in the last year that was both unplanned and unscheduled? Do you know? Can you assume or infer certain levels?

One of the difficulties is in calculating an ROI when there is no information or maintenance history. Some reading this will find this improbable; however, I assure you that it is not only probable but common.

Once you have defined the percentage of work that was done in the past year, you can assume a 50% reduction in its cost, without any further problems. Again, the implementation of a CMMS is not only a change in what you do but in the way you do it and the underlying philosophy of how you do it.

So what else? What about the indirect costs associated with machine downtime and poor reliability. A CMMS enables you to analyze this information in detail to determine the root causes of machine failures. Although RCFA is an area requiring very specialized focus, there will be failures that come to light with available data that would not have been noticed normally.

It is difficult to state exactly what the effects of a CMMS can be on availability; however, from my personal experience a conservative estimate is 2 to 3% within the first 6 months of operation.

What would an increased availability of 3% give your operations? If applied to a fleet of haul trucks, it may mean the difference between 10 and 11 trucks. This means you can rest some of your capital, perhaps even sell it, as well as the flow in savings to maintenance requirements and inventory holdings. Obviously this figure must be analyzed on a per-application basis, but it is always a significant savings.

What about reliability? By having the data available to analyze, you may be able to raise reliability easily by designing out or developing better maintenance procedures for nuisance faults or trips. The savings here can be 100% when the repeat effects of a small trip are calculated for impact on operations and the resources carried to manage it currently. The issues of availability and reliability will be discussed in the section on reporting.

The last, but by no means the least, area of savings is that of inventory levels and control. Most modern CMMSs provide the capability to report on slow-moving items, to classify the inventory holdings, and even provide optimization in some systems. When added to the flow in inventory savings from increased maintenance efficiency and machine availability, this area may well reflect the greatest dollar savings available within a short time frame.

Also within the area of inventory management, introducing systems with the capabilities to manage inventories in an advanced manner opens the possibility of the organization adopting such methods. For example, an organization may suddenly have the tools at hand to manage consignment stock and vendor-held stock, without mentioning the possibilities of linking to Internet marketplaces and even further reducing stock holdings. Personally, I have seen savings of 5 to 7%, conservatively, in inventory holdings, mainly due to the application of advanced reorder and quantity algorithms. As discussed later, these are vital to consider and may well determine the type of CMMS your organization requires.

It is important to note that the figures I am quoting here are very conservative and not reflective of organizations where there is a significant effort dedicated to cultural change of the workforce in order to fully adopt the principles and practices of advanced maintenance management. Returns on investment very much greater than these are not only possible but frequently achieved.

Once the probable savings are calculated, then the budget limits for the purchase and implementation can be set. At this point, you can determine what your own annual ROI will be. For example, if you are looking at a savings of $2 million a year on maintenance, inventory, and indirect costs, a budget of $20 million will yield a 10% ROI. As can be seen, the figures and size of investment here are not for the faint-hearted, nor is it a matter to be rushed or entered into lightly. And, as always, it will require a great amount of leadership, effort, and focus to achieve the goals initially developed.

It is not advisable to allow the consultancy or software firm to calculate the possible ROI. Even with performance-based contracts, there are a multitude of possible factors affecting such calculations, and it is best that your people are definitely involved in, if not leading, the final ROI calculations.

So, a case can easily be built for the purchase and implementation of a CMMS. Now comes the need to define exactly what it is that you want. The definition of business requirements and rules, while vital for CMMS selection, is also a core element toward the advancement of the organization to the planned state of maintenance management. And while the requirements of the business are addressed here, the business rules as they apply to each process will be detailed later in the book.

Business requirements

We begin with the selection process. You now have to determine the requirements of your business that you want to manage. The more prepared you are in this aspect, the more control you will have over the entire selection process, and you will be better positioned to evaluate all of the proposals in terms of the functionality required.

Much of the tendering process is generally governed by determining what it is exactly that a client wants. Although clients often have a very good and detailed understanding of what is required, when faced with the array of providers and functionalities available, the requirements grow considerably, as does the time to quote the implementation due to adjustments to the changing requirements.

It is essential to determine what the size of your asset register truly is. How many items of equipment do you need the system to manage and to what level of detail? Some key elements to consider are discussed next.

Equipment Details

- What exactly are the equipment details and what is critical in the list of things you need to manage?
- Is the equipment register kept in a structured format? Is there a standard for numbering the equipment?
- If you have a numbering structure, how do you want to report costs?
- Do you want to trace equipment under warranty, and treat it differently?
- Is your equipment easily classified into groupings? If so, do you want to do this within a system? Often the ability to group the equipment will facilitate easier entry of statistics and component structures for materials management.

Equipment statistics

- What are the statistics that you want to follow? Some examples here may be:
  - Tonnes, liters, or other method of recording production
  - Hours worked, journeys done, or number of times the process is enacted
  - Do you want to know unit costs of the equipment? Or is it adequate to know the unit costs at a higher level within your numbering structure?
- Do you want your CMMS to tell the expected life remaining on a piece of equipment?
- Do you want to use the equipment statistics to determine your maintenance frequencies?

Equipment costs and history

- What are your requirements in terms of reporting? For example, you want to know the costs of the equipment for:
  - Activity
  - Saleable or total units produced or processed
  - Accounting period
  - Profitability
- In terms of history, you need to be thinking about what is required from the system. Do you want to record only the activities that have work orders associated? Or do you want to look over the entire history of the equipment?
- Related to history and costing is the ability to perform equipment and component tracing. This is an exceptional function that I would recommend to any prospective buyer. The concept here is to trace vital or costly items from their purchase through to their ultimate end of life or sale. The value of this sort of information is powerful in terms of warranty or comparative supplier analysis.

Business rules and guides

This is among the more important steps that will be required to take you to the planned maintenance state and set your organization up for the journey to the proactive maintenance state. During this step of the process, you can look closely at the business rule required for each process and procedure to function correctly. Business rules include but are not restricted to:

- Role descriptions and profiles of candidates for each role within the maintenance organization
- Levels of authorization and responsibilities of each role
- Definition of what constitutes a modification or capital funds project
- Determination of priorities of work
- Determination of what qualifies for a work order and what does not
- Policy for inventory management
- Determination of codes used to identify maintenance types
- Determination of the work-order life cycle
- Definition of critical human resources rules, such as criteria for employment for roles, career planning, and other areas (the most critical and at times the most expensive costs associated with maintenance management are often the human resources element; hence, there is a need to include a focus on this in the work of defining the maintenance requirements)

The rules that are created at this point in the process will determine the final outcome of the transfer of your organization from the reactive to the planned state of maintenance. Lack of focus or errors in application in determining the business rules will result in systems that are either ineffective or are not durable.

More focus will be given to this theme as it applies to each of the subjects that are raised during the course of the book.

Business processes

This item will be covered in greater detail in subsequent chapters; however, at this point it is vital to recognize the processes that your organization has in place, and which of them you want to carry into the new system. It is recognized that a system will generally give you options to a greater range of processes than can be applied. Following is a list of areas that are worth considering.

- What are the purchasing procedures utilized by the organization, the limits of authorization, and who does what in the process?
- Do you have specialized processes for inventory management?
- Are there specialized planning and scheduling processes?
- Are there capital works and modifications processes?
- Is there a requirement to manage shutdowns or programmed maintenance outages?
- What are your production processes? Do you want the ability to plan and conduct maintenance in conjunction with production, or do you want to manage it separately?
- What are your processes now for managing priorities? Is there a need for automatic escalation of priorities?
- What are your processes now for managing backlog?
- How do you want to plan work for the future? Do you see a need for automatic planning and scheduling of tasks?
- Do you have one site or many? What are the interrelations, if any?
- Do you manage one group of workers or many? Within these groups of workers, are there distinct groupings of discipline or activity?
- Do you currently record employee time and the level of availability of the workforce? How? If not, do you want to?
- How do you manage employee leave applications?
- Do you want to know what skills and training an employee has, and what level you want the employee to achieve, according to organizational requirements?
- What level of autonomy do you require when it comes to writing and producing reports, and how should they appear?
- Do you want the ability to review online graphics of various critical factors?

A general description of the business will also prove useful in eliminating pretenders in the future tendering process. You are beginning now to

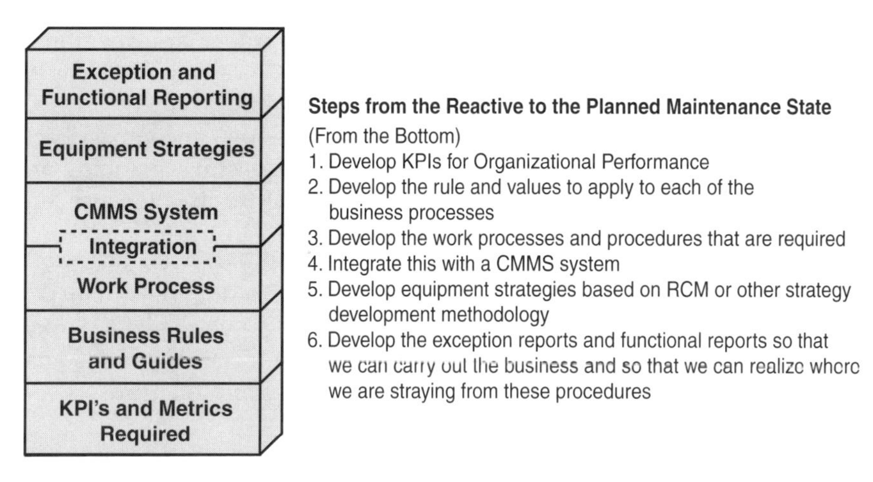

Figure 2.1 Steps to the planned state of maintenance management.

understand what you want from a CMMS. From this point, you are able to put forward a very specific list of requirements to potential providers. You can also begin to look closely at the requirements of your business and see opportunities for change and improvement. For example, perhaps at this point there is no means of managing technical plant modifications as distinct from general maintenance. You can use this time to ask if you need to change your processes and how exactly you should do so.

Although there is no doubt that there are consultants and consultancy companies that are able to guide you in this respect, it should always be remembered that you know and understand your business best. You are best positioned to suggest, plan, and implement improvements, perhaps with a little orientation or training. When looking at your requirements, you can begin by revising the steps required to take your organization from the reactive to the planned state of maintenance, which, along with the business description you have put together, will put you in a very good position.

Quite often when evaluating business requirements, potential clients may not have a clear picture of exactly where it is that they want to be with the implementation. I have found that by using the graphic depicted in Figure 2.1 as a guide, clients have been able to define their requirements easily and quickly.

With the ground rules laid, there can be an earnest effort to develop the business processes that will be required for your developing maintenance department. You need to review and develop all of the processes required in order to carry out the maintenance function in the most efficient and cost-effective manner. It is important that cost effective does not mean only direct costs but also the indirect costs spoken about previously.

In beginning this process, it is essential to have as a base the fact that there are only four processes in the world of maintenance management.

Within each of these lies various procedures which also need to be defined. The four processes governing maintenance management are operational maintenance, technical change management (modifications and capital spending), shutdown management, and continuous improvement.

There are an array of functions within these areas that encompass backlog management, with its sub-functions of work requesting, planning and scheduling, execution management, capital approval for maintenance, and the inventory management array of procedures.

At this point, it is important not to overlook the processes relating to human resources management, as these will determine the quality of resources available and their states of motivation. Here you need to look at recruitment, workforce planning, career management, training needs analysis, skills matrices, and a host of other processes needed to ensure a good-quality workforce.

KPI or metrics development

A company will usually have a range of key performance indicators (KPIs) that it likes or needs to use, or it may believe it can add value when KPIs are applied to the organization, based on the experience of other companies. As with anything, there are many different approaches possible; however, the most useful is to develop a hierarchical structure, possibly based on a balanced-scorecard style approach. Whatever approach is used, it is important that there is consistency and a defined set of measures that you require the CMMS to deliver to you.

CMMS

With the work of defining the requirements via business descriptions, KPIs, rules, and processes completed, the CMMS can now be selected and implemented. Although there are other considerations that must be thought of within this area, the step of purchasing and implementing the system lies at this level of development.

Of course, with the preliminary work of defining rules and processes done, the work of selecting the system now becomes a lot easier. You now know what you are looking for, so when you finally receive the CMMS online, you have already thought about exactly what it is needed for. In other words, you buy the system that best suits your vision of the maintenance management environment that you want to achieve. This will ensure that you have the integration between the processes and the CMMS. It is also important to note that the processes may need to be tinkered with and adapted to suit the CMMS that is best suited to your needs. As long as you are not changing the fundamental vision of where you are required to go, this is acceptable.

Equipment strategies

With the processes and the tools to handle them now selected and undergoing implementation, you are free to look at the content of the CMMS. This involves reviewing the equipment strategies and how best to manage your equipment asset register. This point is not one that I intend to elaborate on within this book. There are scores of systems, methodologies, and variations of these methodologies to choose from. However, at this point, recall that we are talking about equipment strategies, not operating strategies. As such, it is still too early in the development of the maintenance organization for you to be looking at the possible implementation of a TPM-style process.

Functional and exception reporting

The last of the series of steps that you need to develop, implement, and embed is the range of reports that your personnel will need to do their jobs correctly.

Functional reports are required to do your work directly. For example, a functional report can track all of the work orders that are planned, sorted by priority rankings, or a report on the work orders or work requests that have been raised during the past week so that we can audit them, correct coding, and begin the planning process.

Exception reports are those that enable you to easily determine where your system is failing and not functioning properly. An example is the age of work orders by priority. As we will see later, priority must be correlated with time. As such, you are able to find those work orders that are not being managed in due time. Another example is a report of all of the work orders with codes that are not congruent with your business rules. This also will be further explained.

Project management

We now come to the area of setting in place the management team that will manage the entire implementation from the definition, selection, and purchase of the system to the go-live after the implementation phase. This team is of critical importance because they will define the way in which you do business for the foreseeable future. They will need to be fully empowered and trusted to make decisions of a critical nature regarding budgets, time frames, and any variations to the original plan. This will depend greatly on the number of sites being implemented at any one time, but, basically, the structure will be as follows:

- *Project director:* Responsible for the entire project and its eventual outcome.
- *Project manager(s):* Depending on the size of the project, there may be a need for more than one PM, possibly one for each site.

- *Functional team leaders:* The people responsible for their operations area. In a large-scale implementation, there is a need for a leader in each of five areas:
    - Maintenance
    - Finances
    - Human resources
    - Operations
    - Supply or store functions
- *Key users*: In short, these people are the guardians of the new system. They are responsible for advising the incoming consultants of the decisions made with regard to business rules and processes; for identifying information that must be transferred from the legacy systems; and for relaying the training they receive to the end users during the implementation.
- *Technical staff*: It is always advisable to have a capable group of technical staff on hand to learn about the new system, prepare reports, and act as troubleshooters for the functional staff. Of course, the learning curve for these people will be large; however, it is necessary that they start in the process as soon as possible to reduce future reliance on the consulting team.

At this stage, it is still too early to consider the implementation plan; however, an overall time frame can be envisaged. The system that is eventually chosen will come with its own methodology and implementation process that can be adapted to or changed by your staff.

The last items that must be considered when undertaking the functional requirements of the new system are interfacing and migration of data from existing systems. Prior to engaging a software firm or a consulting company, it is wise to have a good idea of the existing systems that will be interfaced with the new system. There is no need to go into the detail of how this will be done; that is the consultant's job. However, it is critical to realize that it needs to be done.

Some examples of legacy system interfaces may be within the following areas:

- An existing RCM or other expert-type system
- A root-cause analysis system
- An operational management system, such as a PLC or other process management system
- An inventory optimizing package or criticality analysis package

As with migration of information, it is wise to review exactly what history is required to transfer from an existing system to the new system. This will afford the consulting team additional time to do the job correctly and ensure there is no loss of vital data.

At this point, you are 90% ready to begin your selection process. Subsequent chapters further explain how you need to go about the work detailed here. With these subsequent chapters, you are truly able to prepare your implementation template in a manner that can be implemented at any site within your operations in a fraction of the time normally required, thus cutting substantially the up-front cost and increasing the potential return on investment.

chapter three

The key areas of maintenance

Reviewing the business rules and processes for the maintenance function of any organization can easily be a very major task in which the level of complexity can become very confusing, very quickly. It is essential to realize at this point that in addition to the hidden element of human resources planning and management, the operations of any maintenance department basically fall into the following main areas:

- Operational maintenance
- Technical change management
- Shutdown or maintenance outage planning

Within these areas are all of the processes that can possibly fall under the heading of maintenance management. This can include apparently disparate processes such as capital acquisitions, cost controls, and equipment strategy reviews and updates.

Running through these maintenance areas are a series of common processes such as backlog management, planning and scheduling, execution, and analysis or continuous improvement initiatives. How these common processes are to be managed will vary, depending on the area in which you are managing them.

The idea of this chapter is to direct the focus of your analysis to the subprocesses of each of these areas and analyze some of the business rules that are needed to manage them.

Operational maintenance

The definition of the operational maintenance area is the area that applies to the day-to-day maintenance of the plant and the organizational interfaces that are required in order to do so. You need to analyze this area in fine detail with the objective of highlighting and defining exactly what it is that you will be requiring in the way of rules for your business.

Outsourcing

Before looking any further at specific areas, you need to be asking yourself exactly what it is that your people are going to be doing, and what tasks or additional work are you able to contract out to other organizations.

There has been much advancement in the area of outsourcing for maintenance requirements but there also have been many failures and opportunities to learn in this area. As such, there are some general assumptions that can be made regarding the area of outsourcing. First, what is offered by contractors as benefits to you for outsourcing work to them? With the benefit of hindsight, it can be seen that many of these reasons were not strongly justifiable in the beginning and are even less so now. Although I am against the practice of piecemeal outsourcing in general, I feel that there are conditions and situations where it is not only necessary but a good investment. A classic example may be the outsourcing of highly specialized tasks that are seldom or infrequently done.

Some may say that this is a good case for increased training of employees to do the work. However, as the task is done infrequently, the level of skill needed may not be adequate and, when an employee receives additional training in a specialized area, he becomes more valuable in the marketplace. There is the possibility that you are training the employee for somebody else to use.

The practice of outsourcing complete maintenance departments, however, is a different area totally. There are some very strong and valid arguments for outsourcing. When linked to KPIs (key performance indicators) or operational concerns, they can make a strong case for cost effectiveness and added motivation to improve overall maintenance performance. However, why this interest at this point of the book? Because it greatly affects the control that you will require your CMMS to manage. If you are going to outsource your maintenance or parts thereof, you need to define exactly how you are going to do it.

If consultants are to be used as an additional part of the labor force when workloads require it, there needs to be the capability to manage this; if they are to be managed and paid via a contract arrangement of some nature, this also must be entered, controlled, and even scheduled and planned where required. In addition, what will happen with materials and parts that are required by the contractor?

There are, of course, many variations of how to manage these situations. At the outset, you need to be sure of exactly how you want to manage the issue, or to be aware of the possible variations of managing it. This will form another vital part of your request for tender from the CMMS marketplace.

Among the approaches that I have seen used successfully are direct contracts, and creating a work team of contract workers where they can have work orders created for them, solicit parts from the store system, and have all of the costs accorded to their rightful place via cost elements and cost codes.

The temptation to give contractors access to the CMMS in order to program their own work should be resisted, unless the contractors are closely supervised. A work order on a site where there are sub-contractors is basically a contract for work, and this can lead to cost overruns and logistics issues.

Work order and work request creation

It should be noted here that I am speaking only in terms of corrective actions. This is due to the fact that as much as 90% of CMMSs on the market today have some form of functionality to automatically create work orders for repeating tasks, which generally fall under the headings of preventative or predictive condition-based maintenance tasks.

What are the types of work that can be done via work orders, and how are work orders created? There are a few areas that you can focus on. Work order creation for maintenance purposes falls into four broad areas:

1. Those arising from preventative inspections or predictive maintenance tests
2. Those arising from operational rounds and inspections
3. Those arising in the form of breakdowns during the operation of your plant or equipment
4. Those arising from equipment maintenance requirements that cannot be programmed automatically

Each of these items needs to be defined in terms of who raises each one, and the level of detail you will be requiring. In the case of a maintenance practitioner creating the work order, you should expect a greater level of detail than expected from an operational person. This is due basically to the maintenance person's familiarity with the processes and requirements. Both are familiar with the equipment and will be able to raise a work order or request in substantial detail from their unique perspectives.

An exception may be the planning and work order creation required for a shutdown or period of maintenance downtime that is not scheduled on a regular basis, i.e., based on calendar hours or equipment statistics. This will be dealt with later in the chapter.

Defining the business rules in this respect is extremely important to your continued operations. Why? Because you are determining the criteria that you want to apply to creation of a work order, and how you are going to execute this function (by defining a process).

The criteria that you have for controlling what will become a work order are critical because they will determine what will occupy your planning, scheduling, and executing resources, and that will affect how your plant operates. If you define your criteria in a way that ensures that any work order is focused on the overall performance of the plant and its improvement, you can be sure that all of the material and human resources that are available will be focused into areas where they can be utilized most efficiently.

I have included some basic disqualifying criteria that can be applied to any situation, plant, or location:

1. Is this work needed to maintain the plant operating at or near current levels of capacity?
2. Is this work needed to maintain the safe operation of your plant or equipment at current levels?
3. Will this task increase the productivity of the plant or equipment?
4. Will this task increase current safety level of operation for people or equipment in your plant?

Taking the first point, how do you define the rules regarding work order and work request creation when the task arises from a maintenance inspection or service? Who will create this and how will it be created?

There are two distinct forms that you can utilize, depending on how you want your maintenance operations to function. In addition, there are some issues that you need to focus on, in terms of the use of maintenance time. You can use a work request system, or you can use a work order system.

The work request system is very useful, and is essential for control over the quality of your backlog system. You can apply the work request system to tasks raised from maintenance tasks, and by maintenance personnel, if required. This will give you added control by limiting the people who determine what becomes a part of your backlog system.

Also, part of the control function that you need is to ensure that planning time is not wasted, and maintenance efforts are adequately focused on trying to limit or terminate the following situations:

1. Multiple work orders for the same task
2. Work order creation with little or no useful data, e.g., "it broke"
3. Maintenance of beneficial tasks (to be explained in more detail)

Using the work request system, you can limit the function of work order creation to maintenance planners and superintendents, and leave them to do the auditing and reviewing of such work requests to ensure that you have the coding required to initiate the work order life cycle. The process in this instance is:

1. Service or inspection carried out
2. Work requests raised by maintenance personnel
3. Work requests processed and raised as work orders, where applicable, by the maintenance planning or management staff
4. Work orders, related to the original work orders via field links or other database mechanism, assigned to individuals or work teams for execution

Can you use maintenance practitioners to raise work orders directly? I feel that the answer is yes, these individuals probably can be used to raise work orders. However, if this decision is made, there needs to be an additional element of control in the process, i.e., exception reporting.

Why is control a valid issue here? First, as mentioned, it is important to control the work orders that are created in order to better control the ultimate quality of work done. Second, you need to control what information goes onto a work order as a minimum during the creation phase so that you are able to cut down the work of the planning department and allow for ease of work order management.

The person creating a work order should be the best qualified to state what work is required, and what resources may be required to carry it out. Where possible, the following information should be mandatory or preferred as part of the work order creation process:

- Description or equipment number
- Full description (or the best possible) of the fault noted and suggested work required
- Name of the person raising the work order
- Human resources estimates
- Materials estimates
- Priority as defined by the work order creator
- Correct maintenance definition codes
- Codes to determine whether this work is best covered by a maintenance outage or shutdown

The exception reports mentioned previously must be designed with the ability to detect work orders that are lacking in any of these points of information or have coding structures that do not comply with the business rules.

As a part of process definition, there must be a critical review of the number of craftsmen that will be employed in the maintenance areas and a determination as to whether they are really the best people to raise work orders. Although it is very easy to get wrapped up in some of the advanced functionality that abounds today, it is also worth remembering that the maintenance department does not exist to feed information into the CMMS; it exists to carry out maintenance functions. Care must be taken to ensure that the primary focus of maintenance execution is not lost.

An option of creation or entry by maintenance staff is to utilize a maintenance clerk to carry out the task of work request or work order creation via information supplied from the workforce on paper. This can be vetted by exception reporting, planner activity, or both.

In points 2 and 3, you need to look at the process by which maintenance interacts or is interfaced with the operations arm of the business. I feel that it is mandatory that you include the work request process. There is generally some resentment to this from the operations departments, particularly in more authoritarian organizational structures. This is due to the fact that they see this as a loss of control over their equipment or processes. To a degree, they are correct in viewing it as such; however, as previously stated, quality controls must be exerted over the information entering the work backlog system.

Although we, as maintenance practitioners, would like it to be so, the world and our businesses do not revolve around the maintenance requirements (more's the pity); the equipment, the plant, and processes, in truth, do belong to the operational departments. However, it needs to be recognized, and this can be a difficult point to make, that the reliability of the equipment you operate is the responsibility of the maintenance department, and accountability measures are needed to ensure that this is very definitely the case.

Now that you have taken away the ability of operations to create work orders, how do you replace that. Here is where consultants have a golden opportunity for collaboration with their clients. One of the important steps to take is in the daily and weekly scheduling functions; on a daily basis, consultants want to be able to sit with the operations department, possibly at superintendent or planner level, and work through an agenda consisting of four parts:

1. Revision of emergency work orders for the past 24 hours with a focus on quick understanding of possible root causes and possible action to eliminate these problems. Obviously, the issue is that you want to eradicate 90% of emergency work orders, particularly chronic or nuisance trips or faults.
2. Revision of work requests for the past 24 hours, and agreement on which should be converted to work orders.
3. Revision of work orders for the past 24 hours, and agreement on an adequate priority ranking.
4. Revision of the weekly plan as it applies to the work to be done today. Changes should be made openly in order to accommodate changes in priority of requests and work orders from the past 24 hours.

In this instance, not only have you been able to guard your backlog management system, but you also have been able to form a bridge to the operations department in order to better manage and process the work of the various departments on a daily basis — a truly collaborative effort in work scheduling and prioritization.

On a weekly basis you need to have your scheduling meeting prior to committing resources to the following week's activities. This is generally a standard activity these days; however, in the context of the weekly meeting, it may be driven as a more collaborative effort once operations finds itself involved in the day-to-day running of the maintenance decision-making process.

All of these requirements point to more details that you require of your CMMS and an even greater level of functionality that it will be required to deliver. However, there is a point that you need to consider that does not come under the headings of work request or work order raising procedures: emergency work orders. When you have an emergency, there are two things that occur:

1. You need to do the work first, rather than raise the work order.
2. You often need items from the store; to requisition these electronically often requires an open work order. The alternative is to bypass the system manually.

Obviously, in cases of extreme emergency there is a need to do the work first. Once you get to the point where stored items are needed, you then should raise a work order through a relevantly authorized person, possibly even an operations supervisor in the case of a night-shift emergency.

This must be well managed to ensure that work orders are raised for critical breakdowns only and not for whatever takes someone's fancy. A possibility may be training in work order raising criteria and coding, as well as some form of exception reporting to ensure that the person is operating the system correctly.

The last of the work order raising requirement generally applies to shutdown or maintenance outage periods and will be dealt with in the discussion on those topics.

We have analyzed the rules or business structures pertaining to who does the work and how the work order life cycle can be initiated. As always, the decisions need to be recorded and prepared for the request for CMMS tender documents.

There are, of course, many other areas where you need to set the rules for how you are going to go about your daily workload in maintenance management. One of the main areas that you need to deal with is the coding of your various work types and priorities. Although it is agreed that this is a business rule and should be treated as such, there is an entire chapter devoted to codes and other areas of standardization within the system.

Technical change management

Technical change management is often the second most important area of maintenance management. Within this area is a range of processes and functions such as technical modifications to plant, system changes in the case of PLC-type control centers, capital acquisitions of new equipment, and changes to operational procedures.

There are two basic reasons why this process and set of business rules are required elements of any plant. First, there is always a need to retain control over all work happening under the control of the maintenance department; second, there is a definite need to record, track, and monitor any changes to operational conditions in order to evaluate and ensure that the technical integrity of the plant is not compromised in any way.

At all times you need to be confident and possibly to be able to prove that the technical integrity of the plant, both in terms of its structural soundness and its modes of operation, has in no way been compromised so as to affect safety of workers or equipment and environmental soundness, and does not present a change in its relation to nearby communities (if such a situation exists).

There are both legal and moral obligations to ensure that this is the case. There have been many cases where engineered change to equipment has resulted in fatalities. We do not want the plant to be open to prosecution or legal action; nor do we, as responsible corporate citizens, want to wittingly or unwittingly cause harm to workmates, surrounding population, or surrounding environment.

One of the key elements to managing this area competently is to define what constitutes a technical change in the plant, and then define how the process affecting this differs from day-to-day operational maintenance procedures.

First, it is necessary to look at what will constitute a technical change. A great many of the ideas for a technical change will come generally unsolicited from the operations environment by way of suggestions to correct a reoccurring problem or to safeguard against possible contingencies that have been noted.

Another source will be management, as a by-product of other engineering works, a direction from the maintenance or operations manager or a direction from the corporate level (normally in terms of capital equipment acquisitions). They also will come as reactionary measures to incidents or accidents in the operational environment. Of course, this is regrettable and any effort to analyze and eliminate problems of this nature before they occur is always the better option. However, the reality is that this will be the source of some of the technical changes that are required.

As a primary step, it is wise to have all incoming work orders processed in a similar manner via the work requesting system in operation in your plant. At this point, they need to be sorted and put into their prospective process streams as required. The criteria to define a technical change request will vary from plant to plant; however, the following may apply as a general rule:

1. Does the request involve a change to the plant's P&ID (piping and instrumentation drawings)? This is often quite an encompassing rule as it will capture anything from a recommended change to a gasket or a valve right up to major structural engineering changes.
2. Can the request affect the operational practices of the plant? Here you are concerned with ensuring that no changes to operational procedures nor anything that may cause changes to operational procedures

can go unnoticed. Again, although to the people involved a slight change in operational procedure may not mean much, however, when reviewed by all concerned with the plant's operation, there may well be side effects or delayed effects on machinery, safety, or environmental concerns.

3. Does it differ from operational maintenance concerns? The defining characteristic is the definition of "like for like." So when there is a replacement of a component for a currently unknown or unaccepted substitute, there is a need to cover this in the technical change management process.

With these definitions, you can apply the selection criteria to determine which of the work requests raised fall under the category of technical changes. There needs to be some form of ability of the CMMS to manage this initial step. The remainder of the technical change process must be defined in association with the organizational structures in place, as well as the levels of authorizations and responsibilities.

It is also valid to note that there are often separate departments in large organizations to manage maintenance and capital projects as separate entities. You should be able to differentiate by using the coding and other processing functions in the CMMS. Whether this is a justified separation will depend very much on the corporate culture and the nature of the business.

One of the key faults of this form of organizational structure is that there is seldom full utilization of the available maintenance resources to carry out the execution of capital projects. This can occur through political divisions, lack of communication, or a perceived lack of skills within the other departments, three situations that can be avoided by either enforcing or implementing strong teamwork principles or communications rules.

Alternatively, this can be overcome by not having separation in this area, but specialization integrated with the overall maintenance focus. The potential savings in terms of spending on contractors can significantly change the ratio of maintenance to operational spending.

Now that you have determined how to define such work requests, one of the important issues is who should define them. In smaller organizations, this will most frequently be a task for the planner in conjunction with the maintenance superintendent of a department and so forth. However, it is wise to include the operational superintendents or managers responsible for particular areas. As such, there can be a unified decision made on the recommendations regarding this project. Depending on price and responsibility levels, there is also the possibility of immediate action for perceived critical items. Applying the criteria with a combination of other criticality or prioritization elements, you can arrive at how these items are to be treated.

Under the heading of technical change management, there is a range of different processes and areas, such as engineering change management, engineering systems changes, and process changes for changes to operating procedures. In order to define these, the following criteria can be used.

1. *Engineering change:* A change to plant or equipment resulting in a change to the P&IDs of the plant or equipment; a change to plant resulting in the addition of new structures or equipment. For example, a new or modified stairwell. Generally done as a result of incidents and accidents, cost revision projects, or as suggestions from operations.
2. *Engineering systems change:* A change to the operational design of PLC systems, hydraulic or pneumatic control devices. These need to be controlled for obvious reasons, mainly due to the fact that everything that can possibly be affected needs to be considered prior to making such a change, particularly in the case of an emergency or alarming function of some sort.

Normally these will come from the same sources mentioned previously; however, they will also come from the systems that you have in operation for managing electrical and mechanical bridges.

Another vital part of the technical change management system is the management of electrical and mechanical bridges or bypasses in the system (related with engineering system changes). Often these are put in place in good faith to enable you to temporarily bypass a nuisance fault or intermittent issue regarding the operation of your plant. You need a way to review and monitor these, as well as a manner in which they can be transferred into an engineering systems change, if so required, in order to finalize the issue and leave the integrity of your operating systems intact.

A final consideration that may be required when considering how to manage the processes associated with administration of technical changes is to provide a fast track authorization and execution path for critical items, either via shortened processes or triggers that escalate the authorization process or bypass parts thereof.

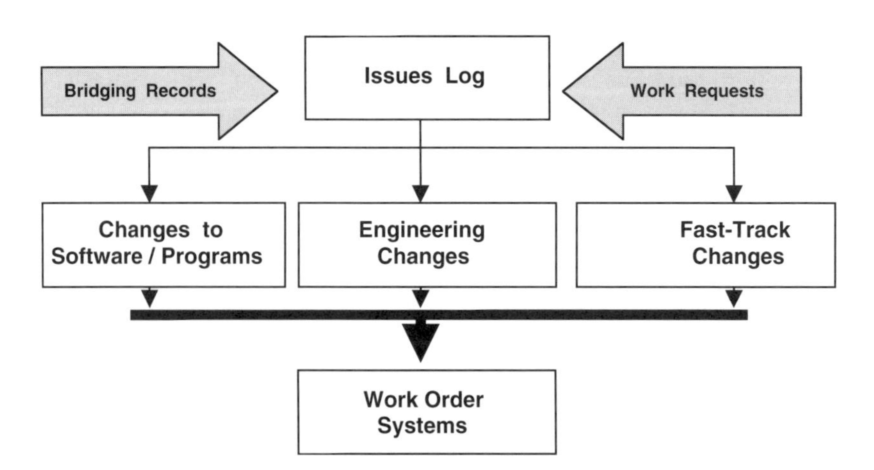

Figure 3.1 Process for technical change management.

Shutdown management

Shutdown or maintenance outage management can often be one of the key areas of control that are required by a CMMS. A great many plants have annual outages for maintenance purposes. These may not be the only instances in which a maintenance outage is required. In some installations, there is a need to effect maintenance shutdowns regularly, e.g., for cleaning purposes or other critical, routine maintenance and operational tasks.

Another area may be a maintenance shutdown in order to respond to a noted fault or a developing fault. These may be required on short notice and the CMMS will need to respond to this in a timely manner.

If an organization utilizes maintenance outage periods or maintenance shutdowns as a part of its overall asset management strategy, there is a very real need to define the processes and business rules that will apply to this area. Everyone who has worked in maintenance outage is very much aware that they are definitely not for the faint of heart. During every shutdown process, there may be additional resources, increased logistical or scheduling demands, a temporarily changed organizational structure, and dramatically increased safety concerns.

In brief, you will have people everywhere doing all sorts of things to equipment in varying stages of operation, depending on the shutdown type, whether it is complete or partial, and operations wanting the plant back as soon as possible to meet their own demands and production goals, etc. Many conflicting requirements need to be coordinated and pulled together in order to have a successful maintenance outage period.

As always, the focus of developing policies and systems of work in this case is to create a workable strategy for tackling shutdowns at your installation, one that is dependent on roles rather than people, and will not disintegrate when a person changes roles. This is a challenging situation and equires the creation of a separate and distinct manual of policies for managing the maintenance turnaround period.

In general terms, you need to define exactly what your requirements are for a maintenance turnaround period. In doing this, you need to keep in mind the requirements of your prime customer, operations. Your focus in this aspect must be on the operational aspects of the plant, not those of maintenance nor those of production, but the requirements of the two key areas.

The requirements of a maintenance turnaround in broad terms are as follows:

1. Safe and efficient plant shutdown
2. Safe timely and efficient plant runup
3. Safe execution of all tasks in the shutdown
4. Shutdown scope to be aimed at the maintenance or reliability requirements of the plant and needs to be restricted to only tasks that require plant shutdown
5. Shutdown to be done within cost and time controls

These are the five main requirements of any shutdown operation. Of course, these will vary from plant to plant, but they are a good base to determine what will be required as part of the shutdown manual. As any system focused in the execution of maintenance work, there will be five main elements:

1. Initiation (or scope creation)
2. Planning
3. Scheduling
4. Execution
5. Review and continuous improvement (including KPIs or metrics)

However, in the specific focus required to execute a successful maintenance shutdown, these will be vastly different from the processes that will be applied to operational maintenance processes, mainly due to the time that may be required for each of them. Through all of these processes, there is a need to stipulate important meetings: who should attend them, and what is the prospective agenda? Although this is an administrative function, there needs to be an ability within the CMMS to place some focus on this. Communication, despite all of the attention it has received of late, remains a very poor area in most companies.

Also, there are various means of executing a shutdown. My contention is that the safest, most cost efficient, and productive form of shutdown management that exists is to utilize the operational maintenance staff to manage the entire process of the maintenance shutdown. Often there is a reliance on contractors to manage this in its entirety, or to manage parts of this process. I disagree with this form of execution as those performing the shutdown should be stakeholders in the process. As such, there is then a need for the shutdown team to be appointed from the operational maintenance teams, and the roles that they will perform during the shutdown must be defined to clearly highlight the responsibilities, levels of accountability, and the time frames in which they will need to carry out each of their responsibilities.

Initiation (scope creation)

This is one of the fundamental blocks of the shutdown process. The formulation of the shutdown scope inclusions process will determine not only what you are intending to do but also what resources will be required to do it. A poorly defined or poorly implemented scope inclusions process can cause a great variety of problems, ranging from poorly directed resources to late or increasingly troublesome startups.

For those directly involved in the maintenance process, how many times have you seen late, unplanned or unauthorized work included in the maintenance shutdown? Many people see this as a time to complete all of the work that they have been wanting done. They will pressure whomever they need to in order to have it done, quite often at the expense of planned and scheduled shutdown works. There is also the risk of safety incidents for things that are "off the plan" and therefore out of the logistics considerations.

There is a need to define in a very clear manner what constitutes a task for a shutdown, the cutoff date for a shutdown task, and the process for task inclusion. This needs to apply to the setting of the original scope as well as inclusion of tasks after the cutoff date and inclusion of tasks during the shutdown. There needs to be clear definition of who does what and when during the task inclusions process so that there can be no unauthorized inclusions and nothing that the management team cannot control or take responsibility for.

What defines a shutdown task is different from site to site; however, the basic rules are that it necessitates a shutdown or it is a high-priority safety issue. If it doesn't require a shutdown, you need to include it only as a low-priority task, i.e., a task that gets attention only when all other work has been completed. Each site must decide its exact specifications, but priority can be determined along the following lines:

1. All shutdown preventative and predictive tasks
2. High-priority corrective tasks
3. Other corrective, preventative, and predictive works on a "nice to do" basis

Scope inclusions processes must be done in three parts as outlined here: first, the setting of the original scope, then the inclusions of tasks after the scope cutoff date, and, finally, the inclusions of tasks.

While using this criteria as a basis for all of the work that must be done as part of the shutdown, you can then create an approval process. Basically, the ultimate authority will come from the shutdown manager, who may be a sectional superintendent or the maintenance manager. But during the shutdown, as previously stated, there will be a need for people to be working dual roles. A lot of the items will be included immediately. Every shutdown has, or should have, a primary reason for needing to happen, and the associated tasks will require only a "rubber stamp." It is important to restate that I am not talking about how to define the tasks via any form of maintenance strategy arrangement. I am merely talking about the process of defining a shutdown scope as a whole.

Figure 3.2 is a diagram of a process that can be used for scope inclusions after the cutoff date. As can be seen, it is a difficult process that relies greatly on justification of items prior to their inclusion. Obviously, part of the reason to warrant inclusion is work must truly be of an important nature. Also there is an effort built into the process to force a more proactive approach to having items included in the shutdown scope.

The final area in which you need to define items that will be included in a maintenance shutdown is in the area of items that are uncovered during

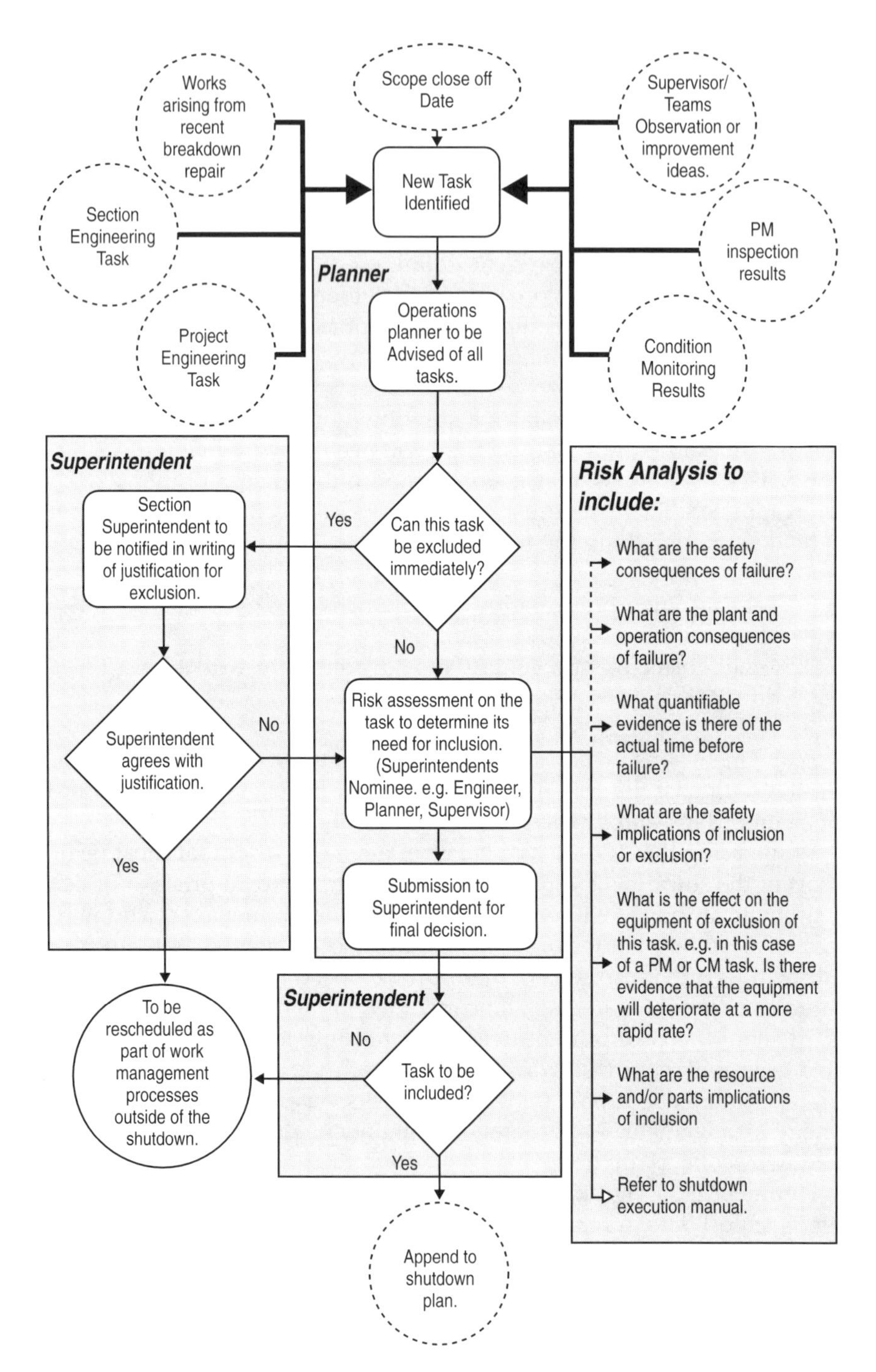

Figure 3.2 Scope preparation for shutdown work.

the maintenance turnaround itself. Some may be due to an inspection or due to an incident during corrective maintenance.

Obviously, during a shutdown there is no time to apply a complicated or difficult process. At this point in time, there may need to be some delegation of decision-making authority to the shutdown supervisors or possibly even to the craft-level workers. This delegation should, as always, have guides to what can and cannot be included, possibly even a rapid form of risk analysis based on Figure 3.2.

Any risk analysis that is performed, using whatever process that is chosen, must account for the following as a minimum:

1. Safety consequences
2. Operational consequences
3. Shutdown consequences
   a. Timeliness
   b. Tasks that will be excluded due to the need to do this
   c. Additional costs of the shutdown

Planning and scheduling

Every task in a maintenance shutdown must be extremely well planned and very well scheduled, ensuring all procedures, materials, and other items are taken care of. The scheduling and logistical side of the shutdown is the main focus.

Pre-shutdown works

Here you are focusing on any scaffolding that may be required, mobilization of contractors and specialized equipment. As part of the safety policies of some companies, there may be a requirement for site-specific inductions or some form of drug and alcohol testing. All these factors need to be considered.

Shutdown execution

The focus is on the day-to-day activities of shutdown execution. Start times, finish and break times, the interrelation of all tasks, and the scheduling of daily communications meetings. This part of the scheduling is the most time-consuming both in the pre-planning stages and in the progress updating stages.

There is a need to include a detailed plan for the rundown and startup of the plant. From this plan, you can begin to schedule the start of various tasks as the machinery becomes available, as well as time limits during startup as the machinery returns to its operational state. Aligned with this is also the need to program in any permits to work or authorizations that will be required prior to the start of any work.

Your CMMS must to be easily updated if there is a shutdown control function. As a part of shutdown scheduling, as it is with construction management, there is always a lot of benefits that can be gained from the generation and frequent updating of an "S" curve. It is a basic measurement device of comparing where you are to where you should be, and, when a project is well planned and scheduled, the plan takes the form of an "S." From this, you are then able to analyze what actions are required in order to fulfill the overall shutdown goals.

Post-shutdown

The post-shutdown scheduling needs to be focused on various areas such as:

- Monitoring tasks for the early parts of a plant's operation
- Equipment and contractor mobilization
- Site cleanup and ensuring that all extra items are returned to the store or discarded (as the need requires)
- Collection of data for the CMMS and its update
- Post-shutdown meetings and review stages

There is often a habit of excluding many of the items mentioned and focusing solely on the physical requirements of the maintenance work.

With the inclusion of all of the tasks in the process, you are able to exercise a greater level of control and, during the review period, you will have a better view of the overall performance of the maintenance efforts. In order to accomplish this, you will need a separate schedule of the pre-shutdown planning process and important meetings.

Figure 3.3 is a typical pre-planning process involving operational and engineering maintenance departments. It is included only as a reference and a guide, and will be site-defined in real-time.

Execution

During a shutdown, it is very important to have people in control of events. For example, the shutdown supervisors are the executors of the shutdown plan; they need to be assigned a group of tasks and a group of workers to do them. This may be by discipline, geographical area, or even by contractor groups (as an overseer). In the schedule, it needs to be made very clear which tasks require which disciplines, and which supervisor will be responsible for the tasks.

From here, with a CMMS that is interfaced with a Gantt charting program or even from the CMMS itself, you are able to give work lists to supervisors, as well as all of the other reporting functionalities associated with this style of information.

During each shift, there should be regular communications meetings. This may even happen daily, depending on the structure of the shift roster

Task Name	
Shutdown Planning Process	70 days
Decision for shutdown taken	0 days
Detailed Scoping	32 days
All PM tasks raised	10 days
Scope Meeting	0 days
Schedulers to code	5 days
Appointment of Shutdown tests	0 days
Detail plan work orders	17 days
Project Engineering Scope	32 days
All works to be highlighted	0 days
Code accordingly	4 days
Scope Cut off	0 days
Handover of initial work scope	0 days
Scope Additions Process	30 days
Final Scope Closure	0 days
First pass Planning	31 days
Shutdown Planning scheduling	14 days
Request extra resources	0 days
Transport Notification	0 days
Craneage to be organized	3 days
Meeting — Handover of Plan	0 days
Detailed Planning	14 days
Final Shutdown Review	0 days
Execution Date	0 days

Figure 3.3 Example of a planning timeline.

in the plant. The information presented at these meetings should be focused on:

- Safety incidents and concerns for the next few work periods
- Operational concerns, e.g., if a task is late and what the effects may be
- Overall shutdown progress and the presentation of daily statistics as support for decision making
- Acknowledgment of exceptional performance, which must be regulated so that it is directed at truly exceptional performance and not merely good or expected performance

The actual execution of a maintenance shutdown can be one of the more difficult and risky endeavors of the maintenance department. It is advisable that the supervision in this aspect is a team of very high experience levels and skills in their fields of expertise and in front-line leadership in general.

Reviews

Quite often I have come across shutdown teams that are totally demoralized. This is due in part to the hard nature of the task but also to the fact that there is no solid evidence about their performance. As such, the hard nature of the role itself can often lead them to be overly harsh in their judgments of their performance. One case in particular was in Western Australia where I came across a team that was quite exceptional; however, due to the lack of hard data supporting this, they judged themselves very harshly on a few anecdotal comments.

The review process of a maintenance shutdown must be formed in two parts. First, there is the anecdotal evidence. What were the comments of the supervisors, planners, and all other key roles involved in the maintenance shutdown? What are the comments from the customers, the operational staff? It is important not to be harsh nor to sugar coat things. Decisions as to the future operations of the process will depend on this, and the comments need to be both objective and focused on improvement. A guide for anecdotal evidence are questions such as: What did you do well? What did you learn? What can you improve?

The second source of data for the review can come from your CMMS and operational systems so that you can really prove what you did well and where you went wrong. These KPIs or metrics can be focused in a series of areas: safety and planning performance, execution performance, and plant performance on its return to operations. There will be more written on this in the chapter covering reports and KPIs.

chapter four

Key maintenance processes

Throughout all of the key areas of maintenance management, there is a need for three main maintenance processes. These will cover all of the requirements of the maintenance function as it applies to each of the three key areas:

1. Backlog management
2. Planning and scheduling
3. Data capture and analysis

Before we begin to discuss these themes, there is a need to discuss an important tool that is rarely developed by maintenance departments but is extremely useful in defining all processes and sub-processes, and the roles and responsibilities that are required to carry them out. This is the work order life cycle.

Figure 4.1 contains a typical work order life cycle as it may exist in any plant. As can be seen here, it also details the roles that are required to take action at various points within the life cycle. This chapter will detail the contents of this example, in terms of the three key processes for maintenance management.

Backlog management

Backlog management is by far the most expansive and critical of the three of these processes. It is through accurate and exact backlog management that you determine not only how various jobs or tasks are handled, but also what work will be done within your plant or on your equipment.

The benefits of an effective backlog management system will impact in various ways on the maintenance processes of your organization. During the course of this chapter, we will focus on each of the aspects of backlog management and how it can benefit the organization as a whole.

Before we can consider how to manage a work order, we need to consider what constitutes a work order for a plant and the processes that go with it. Many organizations still operate without a work request procedure. The dangers of this are great and it can be seen in the contents of their backlog. Backlogs of companies that do not employ a work request system and other type of filtering process usually end up with the following scenarios:

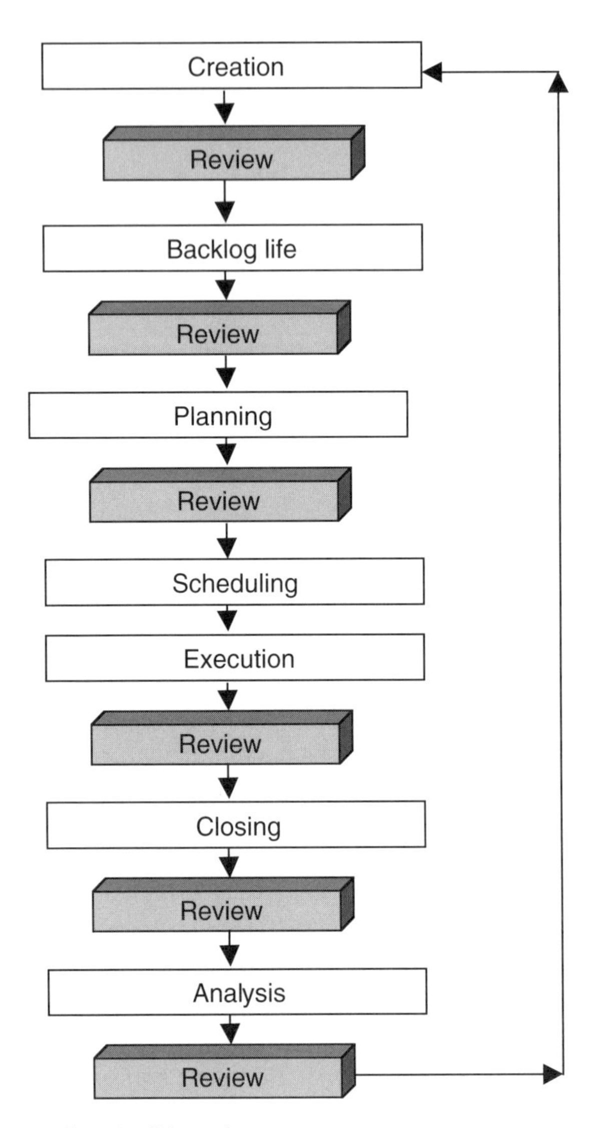

Figure 4.1 The work order life cycle.

- Very many work orders, limiting the manageability of the system. This combined with the errors discussed below can cause problems in planning and programming, as well as execution.
- A large number of duplicate work orders, causing waste of time in reviewing and even working on incorrect work orders. As a maintenance planner, I came across a situation where I was planning three work orders for the same task without realizing it, one of which had already been completed!

- Nuisance work orders. Work orders that normally would not be considered in a plant with a work request system can clutter the system in plants without one.
- A very large number of work orders without sufficient data to allow for planning and execution. The classic example that we all have seen is "it's broken."

So, the intention of the work order system is to establish control over the contents of our backlog management system. As such, we need to define who can and cannot create work orders, and who should be utilizing the work request system to get attention for their tasks.

Work order creation must be restricted in a two-level approach. First, there must be the ability of maintenance supervisors and planners to create work orders. In no way should the imposition of a work request system mean that you are creating bottlenecks in your processes. However, from this point, there must be set rules and policies on who can create a work order. Work requests, on the other hand, should be open to all personnel to create. Through work requests, an operator can register a problem with specific equipment, or recommend design solutions, or register complaints for common faults. Ideal as work order raisers are administrative staff requiring assistance with their facilities.

Having said that, there must be a facility in place for the creation of work orders in the case of emergencies or during the "back" (overnight) shifts.

There is no argument that when there is an emergency situation, it's all hands to the wheel, as it should be. There is often no time to create work orders and deal with the administrative side of things. This is not only acceptable but highly recommended. However, a work order still must be raised at some point for this work, even if it is retroactively. The activities and costs associated with this work order must be correctly allocated and the completed work recorded for future analysis.

On the overnight shifts, there is also a need for special situations regarding the CMMS, perhaps the authorization of someone on that shift to create work orders and order parts. I have seen this as an issue due to the fact that, in many companies these days, there is no night shift supervision. The teams run in a semiautonomous manner.

Once the work request is created, there must be guidelines as to what constitutes a work order. Here I am referring to both the criteria by which we recognize what is a work order, as well as the information that it needs to contain.

As a guide, a work order should be created only if it:

1. Restores the plant or equipment to its original condition
2. Improves the safety of equipment to personnel
3. Improves the operational performance of equipment

From this point, it enters the work order coding and directing phase. Here a role must be defined for who will revise work orders and direct them to their required process. This is most suited to the role of the maintenance planner. As mentioned previously, this must comply with the processes and rules outlined by the organization:

- Operational maintenance
- Technical changes
- Shutdown maintenance

However, prior to being directed, a work order must be coded correctly. Without this step in place, there is the danger of the planner not being able to understand exactly what is required or why and how to later plan and schedule the work order. The coding to be considered here includes:

- The type of work order: maintenance, nonmaintenance, capital, safety, etc.
- The type of maintenance required: preventative, corrective, emergency, etc. (a full and detailed review of the maintenance types and work order types, and how to structure these will be given later in this book)
- The priority of the work order as seen by the person raising it and the desired-by date, if any
- A full description of the fault found and the work required to fix it
- The equipment affected, possibly even to the level of the affected component

Exception reporting involves reports designed to capture information that does not comply with your business rules. That is what we are talking about here: the establishment of rules that you will use to conduct your business in the most effective manner possible. And here we are speaking about the business of maintenance management.

Exception reports will be referred to at various times throughout this chapter and those that follow. They must be applied to every facet of backlog management and the planning/scheduling and data capture functions. Without them, the level of human error or neglect can have dramatic effects on the overall functions of the maintenance departments. An exception report can make the role of a maintenance planner simpler by allowing the planner to quickly extract information regarding the work orders in the system and badly coded work orders can be defined and corrected as quickly as possible.

At this point there is also an opportunity to introduce yet another efficiency tool in the maintenance management arsenal: the 24 hour work order analysis report. The idea here is that each morning the system produces a report, possibly sent by e-mail to the people involved, of all of the work requests and work orders raised in the past 24 hours, listed by priority and by the maintenance type. Ideally, this report is used during morning meetings between production and maintenance departments in order to agree on which work requests will become work orders, and to agree on the effects of additional work orders on the planned and scheduled maintenance work of that day. This will be covered in more detail when we discuss scheduling.

Over time, you are able to measure the progress being made on your maintenance backlog management and on the overall performance of your maintenance department. For example, you can expect, as a result of these style measures, the following types of results:

- Increased ratios of proactive tasks in relation to reactive tasks
- Trending upward of the content of your maintenance backlog that is in a planned state
- Higher percentages of predictive and preventative works
- Higher levels of quality information within the maintenance backlog

Work order planning

In reality, all of the processes discussed in this chapter fall under the umbrella of backlog management. However, as they are all distinct and different parts of backlog management, it is important to view the component parts of each separately.

Planning and scheduling are very often confused within maintenance departments. I regularly see maintenance practitioners refer to scheduling as planning, for example. However, they are two distinct functions of maintenance, each with its own importance with respect to the entire maintenance function.

Some quick definitions: planning is the what, where, how, and why of the process; scheduling is the logistics associated with the "when" of any task. It is that straightforward. Planning is the act of getting a work order, and hence the job itself, to a state where it can be scheduled or executed in an unscheduled manner. In addition to confusion between planning and scheduling, there are some incredible statements in the industry regarding the exact definitions of planning and scheduling. Although they can be described in a very complex manner, it helps to keep it simple.

What do you need to do to say that a work order is planned? This is another company-specific item that must be defined as part of the business rules. Each organization has different rules regarding this, depending on the level of planning that it wants to put into the process. Many times, there can be a tendency for an organization to go overboard and demand a level of planning that either is not cost efficient or just plain impractical.

Listed here are not the "ideal" requirements for a planned work order, merely some of the items that should be considered when setting in place the guides for the future:

- Are all of the required materials available?
- Are specialized tools and equipment needed?
- Are all of the resource man-hour estimates entered?
- Are all of the cost estimates entered?
- Is there a written or required procedure? (Every task should require a written procedure.)
- Are there records of access permits, Material Safety Data Sheets, or any other safety requirements?

When all of the criteria are met, and only then, can it be said that a work order is planned. From the list, what are the expected benefits of having this information or this list of activities completed within the system? The elimination of waste, which is one of the recurring themes in effective maintenance management. The expected benefits are the elimination of time wasting that is involved with:

- Waiting for materials
- Lack of experience
- Not knowing for a certainty what resources are required to do the work (thus creating the possibility that there may not be resources available when the equipment is)
- Waiting time for preparation of access permits or looking for other style documents associated with the work

The other benefits that can be derived from accurate planning information is the amount of planned work that your backlog contains. For a maintenance department to be in a good state of preparedness, it is necessary to have two to three weeks of planned work available.

The question now must be: Why? Maintenance preparedness is one of the fundamental aspects that separates maintenance practitioners from maintenance pretenders. With a strong content of planned work orders in the backlog, there is the ability to react quickly to any number of situations. Following is a list of some of the frequently occurring incidents where maintenance preparation can be of great use.

- When the planned and scheduled work for the week has been completed, there will be a list of planned or "ready to go" work orders for the maintenance staff to choose from. This will be dealt with later under the importance of the planned and scheduled ratios.
- In case of equipment breakdown, there is a bank of planned work orders in which you can better utilize the time with the equipment and the human resources available. This is not to say that the moment a piece of equipment breaks down, you throw all of your resources at it. Your reaction to any event, either planned or unplanned, will always depend on the priorities that you have at the time.

- When a breakdown occurs within a plant environment, there is often associated equipment that is unable to be run, due to process reasons. In this case, you also have a bank of planned work orders that can be executed.

Once an organization has chosen from these criteria what it intends to use as a guide to its planned status, there is a need to know how this will be recorded.

All CMMSs, no matter how advanced, need a planning status indicator field of some kind. This is essential to understanding exactly where each of these work orders is in the planning process.

When there is a concern that an organization is going to have a very large quantity of work orders to plan, there may be a need to look for a CMMS that has the option of automatically updating work orders to indicate when they have or have not been planned. Although different systems have different ways of doing this, the main focus is on materials. Once all of the materials ordered for a work order are available, the planning indicator can be changed to what is the planned state. Again, this is dependent on the requirements of the company. I have seen sites where there are in excess of 20,000 work orders generated daily. Backlog management in such an environment is chaotic, and more often than not, left out, if there were no means of doing so in a semiautomatic manner.

An issue that will be dealt with later in this book is that of work order templates. The primary intention of work order templates is to take much of the work out of the planning process or, in other words, to "bottle" the experience that has been gained over time in the organization.

Now you are going to the heart of backlog management, progressing work orders through their states of readiness until they arrive at the planned or ready-to-execute stages. From this point, your system must be able to filter work orders by priority, then by the planned status.

Returning to the example of shutdown management, you know you have a maintenance outage coming up; you have been through the scoping process and are near the point where you will cut off any further scope inclusions. It is time to begin ensuring, as you will be doing periodically, that the workload is going to be in a planned state.

Using the filtering methods we have listed, you begin your work. What are the materials required? What are their lead times? Will you have them? How can you get the work orders requiring safety instructions or work procedures done? Is there a document relating to this? If so, does your CMMS provide the option of attaching documents (yet another qualifier perhaps)? And so on and so forth until you are able to plan and therefore qualify the work you will be doing during the maintenance outage, and disqualify the work that you will not be able to do, due to long materials lead times and other factors. As always, however, you must be conscious of the priority factors and put additional emphasis on the tasks that are of higher priority.

As a result, you also need to have defined any delay coding that you intend to apply to your work order system. For example, you will need to know if there are any delays due to lack of material or parts, poor procedures, or any of the other items that you should be aware of as part of the concept of having a planned work order.

Capacity scheduling

Capacity scheduling is the most effective way to schedule the time and utilization levels of your human resources. It is a concept based on the maximum resource hours that are available for maintenance work.

Basically, you take the workforce 100% time available and begin to subtract hours from it until you reach the maximum resource hours available. First, you need to take away the hours that you know that employees have off as holidays, vacations, and time allocated to training and meetings. Second, it is important to deduct a percentage of the time available for hours that you need to dedicate to repairing breakdowns. In a perfect world, this will not be necessary and there are some organizations that try to do away with this, calling it a "stretch target" or other similar name. However, if you do not leave an adequate percentage of your time free for the possibilities of breakdowns, you will never achieve 100% schedule compliance. As an ex-maintenance planner, let me assure you that to program maintenance in this way is the surest way to lose the confidence of the supervision staff, leaving the weekly plan open for them to pick and choose the tasks they want to do because "we never finish it in any case."

You are then left with your 100% of resource hours available for maintenance work during the next schedule period. I refer to this as a weekly plan because this is the manner in which many companies do operate their maintenance plans. With the move to the planned state and the subsequent striving for the predictive state of maintenance management, it is possible that there will be higher availabilities and reliabilities expected from your machinery. As such, you may get away with a lesser percentage of resource hours dedicated to breakdowns and a greater level of control over your maintenance environment. At this point, there may be the ability to move to biweekly or monthly plans.

The first maintenance tasks to be included in your maintenance plans are your defined preventative and predictive maintenance tasks. These are the basis of your continued good operations and your move from the reactive state of maintenance. Their intervals will have been, or should have been, very carefully determined in accordance with the failure modes analysis or whatever other maintenance strategy has been applied. The point here is *do not move them!*

Next is the inclusion of planned maintenance tasks into your workload. You will notice that there is very little talk here of whether a task is a shutdown, or an operational or technical change maintenance. This is because they are all treated equally during this phase. It is the prioritization of these tasks, above all else, that determines how they are managed. Note also that here I mention again the use of planned work orders. As discussed earlier, these are the work orders that are available for execution or scheduling.

Starting from your corrective actions, priority one, you begin to fill out the remainder of the resource hours available up to the 100% mark. This is capacity planning and this is how you will realize, within the first week or period that you apply it, that you either have too many workers, too little planned work, or too great a time dedicated to or required by breakdowns. Therefore, at the beginning of the scheduling process, you will require employee work load information from your CMMS based on:

- Holidays
- Coverage for supervision staff (where applicable)
- Training periods
- Meeting schedules
- Maintenance PMs (preventative maintenance tasks)
- Maintenance PDs (predictive maintenance tasks)
- Maintenance corrective actions filterable by priority
- Average reactive workload in resource hours
- Lead times and confirmed arrival dates of parts on order
- Production plan for the period in order to highlight the available maintenance windows and, in accordance with the equipment maintenance requirements, to highlight the periods that you will need to negotiate for.

There are no hard and fast rules regarding this; however, one of the guides should be Figure 4.2. This arrangement of resource hours is typically expected in well organized and controlled maintenance environments.

There is one final part of the scheduling process that should be defined and discussed here: the organization between operations and maintenance. Again, there has always been, and in many cases will continue to be, a somewhat adversarial relationship between the two departments and, despite the fact that 98% of us realize that this is self-defeating and unproductive, it still continues in many organizations. One of the tools that I have successfully utilized to break down the barriers between the two sides is the weekly scheduling meeting. Often this is a regular event within organizations, and it should continue to be so. Once the plan is ready, it must be thrashed out between the two departments to ensure there is agreement on the contents of the plan.

From this weekly meeting, there can be agreements on the dates and times that certain machinery is required. Slight adjustments can be made of certain maintenance preventative and predictive tasks to correspond with the requirements of the operations and a better scheduling of the corrective maintenance actions in accordance with the requirements of operations.

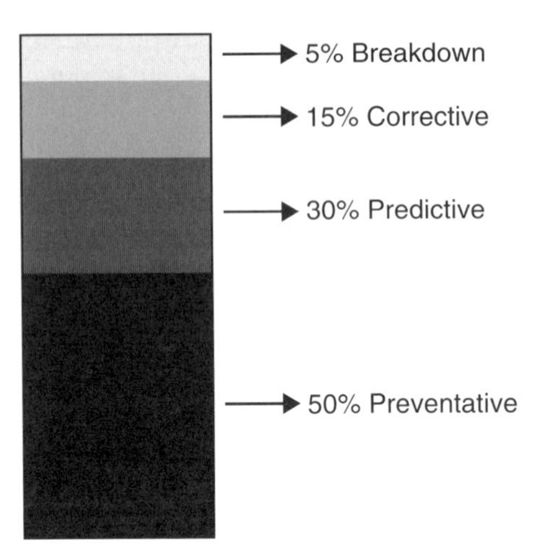

Figure 4.2 World class levels for program content.

It must be noted here that there must be a balance between the operational requirements and the maintenance requirements. Operational departments can seem somewhat shortsighted in these areas. And it can often be difficult to convince them, or prove to them, that the maintenance of equipment is required at this date. With time and the rise in maintenance availabilities and reliabilities of the equipment, there will be a level of confidence in the predictions and concerns of the maintenance department.

Once the weekly plan or schedule has been agreed upon, it can then move to the execution phase. However, there is a need to follow up the plan, sometimes on a daily basis.

There has already been some mention of the daily scheduling meeting between operations and maintenance. There are two main goals of this meeting:

1. As mentioned previously, there is a need to review all of the work requests raised during the past 24 hours. There is also a need to agree on which of these, according to the policies or business rules regarding work order criteria, are relevant work orders and assigning a priority at this point. Therefore, you have an agreement on what work will be done in the plant or on the equipment.
2. The second goal involves the weekly schedule. This must be revised to see how close it is to the reality, where you have gone off schedule and why, and what is the effect of the work orders raised during the past 24 hours, including the recently created work orders from work requests.

Again, according to priorities, there may be a need to include some work orders and exclude others from the weekly schedule. Over time, this will decrease, and the time required for this meeting will also decrease. However, it is a vital tool for the two-party communication and as an assist for each side to understand and take into consideration the situations affecting the other.

Although the majority of systems on the market today are capable of performing capacity scheduling and allowing easy reprogramming of tasks, there is a strong dependence and familiarity with Gantt charting programs. The visual displays therein are easier for many people to understand and manipulate. Therefore, another consideration is, does the system integrate with or contain the corporate Gantt charting software?

Planned/scheduled ratios

One of the more important of the maintenance indicators used to measure the performance of a maintenance department is the planned/scheduled indicators. The planned/scheduled indicators will tell you what percentage of your maintenance man-hours have been dedicated to execution of tasks and in what form. The definition is basically as follows.

While the process of moving a work order to the planned state within a CMMS requires a form of coding, I highly recommend not using any form of coding to state whether a work order has been scheduled. All too often these sorts of codes, particularly the scheduled code, can be "forged," giving a false reading or measurement.

As shown in Figure 4.3, there are many paths that a work order can take. The following list attempts to explain the various paths to execution of a task or work order:

- *Initiate – Plan – Program – Execute – Initiate:* Describes the path from a work request to a work order from its backlog life as a planned work order, through the scheduling phase, and finally to execution. The link from execution to initiation describes the role of continuous improvement of your maintenance analysis systems. For example, there may be changes to procedures, resource requirements, or the initiation of proactive works after an RCFA-style review.
- *Initiate – Program – Execute – Initiate:* This follows the same process; however, there is no planning step. The diagram illustrates the use of work order templates in this path. A work order cannot, or should not, be scheduled without being planned, which defeats the purpose of scheduling.
- *Initiate – Execute – Initiate:* Here you see the typical emergency or breakdown work order life cycle or execution path. In this instance, there is no time for planning or for scheduling, only the execution phase.

Notice that each of the steps in the process overview graphic are backed by functional and performance reports. This will be explained later in the book.

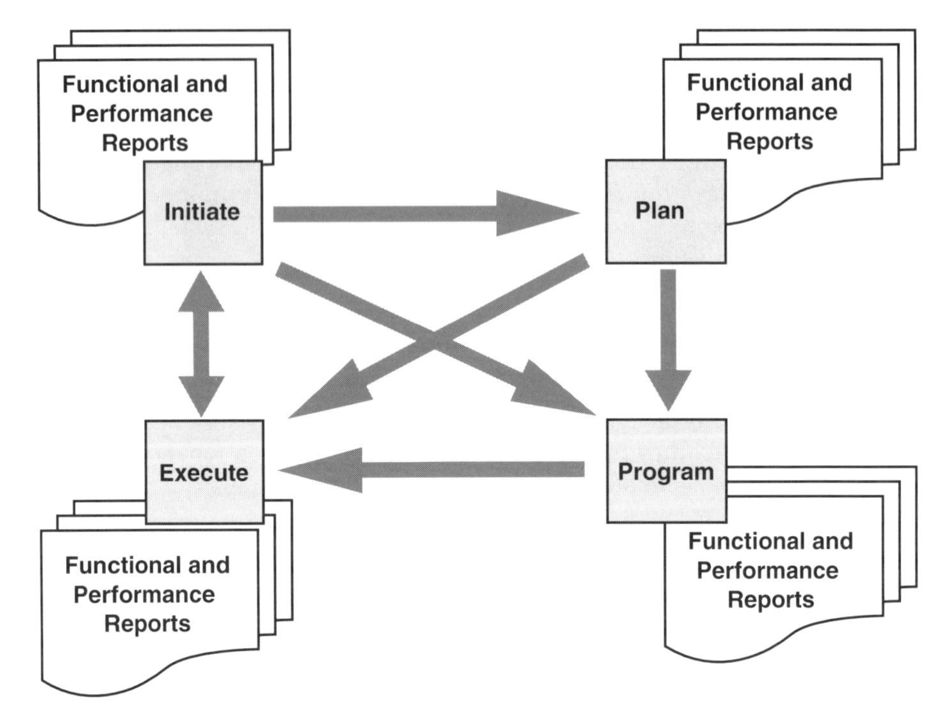

Figure 4.3 The differing paths to work execution.

Planned/scheduled

Work orders that have been planned during their backlog life and have been scheduled as part of the weekly scheduling process provide the best and most cost effective of the three forms of execution.

Planned/unscheduled

Work orders that are executed in this manner are not necessarily a bad thing. A fair content of planned/unscheduled work may show a high level of preparedness and an ability to react to the unexpected and the craftsmen's abilities to self-manage. For example, when they have completed all of the planned and scheduled work in a shorter time frame than expected, they are applying themselves to the bank of planned work orders primarily to ensure the best operational condition of their plant and to keep themselves busy.

The definition within the plant or organization of what constitutes scheduled also affects this area. There are those that will say: "This task came up on Monday; I scheduled it into the plan and completed it Thursday; therefore, it is scheduled." Technically, this is a correct statement. However, you are aiming for the weekly horizon. You want to be in control of your maintenance to the point that what you schedule at the start of the week is done.

There is no need, or very little need, to schedule more activities. This is the true definition of scheduled. As it is not only a form of measure but also representative of a goal to be achieved, anything less is diddling the figures.

Therefore, the bad part of planned/unscheduled is this: when a corrective action occurs that has enough priority to require it being done within the same week but not enough to warrant resources being diverted to it immediately and you have time to plan the work prior to it getting done.

Unplanned/unscheduled

This is what you do not want. The vast majority of these tasks are emergencies and not useful at all. Your goal, as always, is to eliminate the need for emergencies.

It is important to note here that it is possible today to have an emergency task which is planned/unscheduled via the effective use of work order templates. Therefore, this may, in some manner, be a little misleading as to what really happened. However, this remains one of the better forms of measuring the efficiency and control of maintenance departments.

Standing work orders

These need to be mentioned in relation to the planned/scheduled ratios. It is recommended that standing work orders are almost never used for any form of maintenance work directly.

They are dangerous in this aspect because people, like electricity, will always look for the easiest path. As such, when you have a situation where there are no work orders for the craftsmen to book their hours to, they will book them to standing work orders. In addition, when they do not know a work order number, instead of searching for it, they will book their time to a standing work order. The same goes for materials.

Suddenly, a standing work order for resetting a lanyard switch will have 200 hours and $5000 worth of equipment. Standing work orders are never monitored closely enough to ensure that this does not happen. In any case, who really has the time for such things?

Although there are some time issues in raising work orders for each incident and for each occurrence of an event, there is a payoff in terms of the reliability analysis available and the ability to eliminate the task forever. However, there are exceptions.

In the case of a site where there is a genuine effort to capture all of the hours of employees, from operators to maintainers to the janitor, there may be some validity here. For example, with a fleet of 100 trucks and three shifts per day, who has time to produce and manage the 300 pre-start equipment inspection work orders generated daily? The answer is to use a standardized, preprinted procedure and work request generation form and one work order for all. If there is a definite concern for cost accuracy, then possibly one work

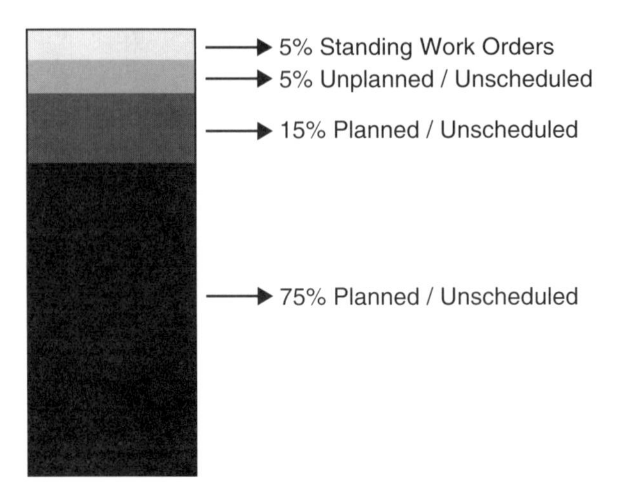

Figure 4.4 World class levels for modes of execution.

order per truck should be used. At least then there can be an understanding of how much time and at what cost the pre-start inspections are carried out.

The genuine use for standing work orders, however, is in capturing hours that are operationally unproductive but are still useful statistics. For example:

- Supervisor coverage
- Training courses
- Meetings, toolbox, safety, and corporate style
- Vacation leave

In this way, the hours spent during the past week also can be reviewed for the content of standing work orders.

What is a reasonable level for each of these components in your analysis of execution of resource hours? The chart in Figure 4.4 provides some indication. The best of the sites that I have worked at as a consultant displayed results such as these, while some of the worst among them always blew out in the area of unplanned and unscheduled execution of resource hours.

You will notice that, unlike the resource hours per maintenance type graphic, this one does include standing work orders. Why is this? Because although the resource hours per maintenance type graphic can be used as a retrospective view for analysis, it is also often used to determine the forecast workload by maintenance type. As such, standing work orders are not required as they are taken out of the equation in one of the first steps of the capacity scheduling process.

Although it may be a difficult area to control, there must be some effort put into it, without lowering the employees' ability to take vacations, to cover for the supervisor as a part of in-house training, and to attend courses to better themselves.

Execution and data capture

Equally important is the process of execution and data capture. There is a need for you to generate quality, accurate data within the system if you are going to have any chance at producing accurate analytical reports that are used for furthering your maintenance development.

One of the proven major causes of CMMS failure is the lack of quality data in the system, to the point that there are often sites or installations where a large percentage of work is done without work orders. In addition, there is a very real need for the craftsmen to receive quality information and procedures from the maintenance planning and scheduling department. As such, this area is a focus not only of what must be done, but of how and why it must be done. And as always, from the definitions in this area, you will be able to highlight further criteria that you will use in your requirements document.

It should be obvious by now that although this is a rapid implementation methodology for CMMSs, it is also a very useful methodology for developing standardized maintenance processes and practices throughout a company. This is part of the concept. With the processes, rules, reports, and general requirements thought out before hand, the actual implementation, particularly if it is a multi-site or multi-company exercise, can be reduced dramatically in terms of costs and time. However, it is not without its pre-implementation efforts and requirements, but we will return to that point later in this book.

Work packaging

The execution stage actually commences with the planning stage, when all of the required elements are put together for execution. Then, once the schedule has been made, the work has been slotted, and all logistics managed, you can begin with the task of preparing work packages.

The intention of a work package is to provide the craftsmen with all of the relevant information they will require to undertake the task in an efficient and safe manner. This may include but is definitely not restricted to:

- Material Safety Data Sheets
- Pre-written permits
- Procedures and safety advice on the task at hand
- Drawings
- Bills of materials
- Copy of the work order itself

Some CMMSs allow for these documents, if they are available electronically, to be attached to the work order so that the work package is already in existence and needs only to be printed.

Taking this a step further, some systems even provide the functionality of attaching these documents to the work order template, so that when it is converted to a work order, the documents already exist. The obvious timesaving functions on the planning side of things are enormous.

Then the work package is delivered to the supervisor for dispatch to the workforce when the task arises on the weekly plan.

An essential part of the work order is the section for comments and work order coding. The coding structure will be discussed at length in another chapter; however, it is important that they are filled out. Without them, you are unable to perform meaningful analysis on your work order histories.

Data capture

Other vital information that must be provided by the workforce includes the following:

- Delays
- Total time taken
- Number of resources on the job
- Materials required to be reaccredited to the maintenance store (if any)
- Comments on the failure and suggestions for future elimination of the failure

The comments that are supplied on the work order are a valuable part of your continuous improvement of work systems. With the correct workplace culture, this can be used to capture ways of doing the work faster, safer, and in a fashion that produces enhanced equipment reliability.

Once the work packages and the maintenance schedule have been handed over to the supervisor, you then need to look at those aspects of data management that are very important to your system, remembering that the goal of these exercises is to define your systems and therefore define your requirements of a CMMS. Again, it is the intention of this document to provide a guide to how you can utilize this period of redefinition to better your maintenance processes and to ensure that you are driving the CMMS selection procedure and that the process is not driving you.

It then falls to the maintenance supervisor to control the work as closely as possible to the workable schedule you have produced. One of the key areas where a maintenance department can fail is either forcing supervisors to do scheduling work, or allowing them to do it. Supervisors will always attempt to do this work. However, as part of the organizational roles that you adopt in this process, you must focus on the fact that a supervisor is a front-line leader. A supervisor's primary focus must be on leading the team to do quality work safely, but with an ever-decreasing requirement of time and resources. This is the supervisor's primary, though not the only, part in your continuous improvement regime.

As a part of the leadership function, the supervisor also must be focused on accurate assignment of work orders to the most-adequate available resource. There are those who believe that assignment of work orders must be done as part of the planning and scheduling process. This is an incorrect approach. During the planning phase, work orders must be assigned to a group of workers or a work team.

However, it is the supervisor who knows the team best, its capabilities and shortfalls, and it is the supervisor who is managing the workload of the team. They are also the ones who must be reacting to various unforseen and unplanned events within the standard work day. This may or may not require them to change the work assignment during the day in order to meet the goal of maintaining production while attending to the regularly scheduled maintenance work. This is the supervisor's interaction with the scheduling function on the hour-to-hour basis where there is a need for decisive action to maintain your assets operating in good order.

The final part of the execution and data capture role that is required from your supervisors is that they ensure that every work order issued during the scheduling period is returned to the planning department. The procedure of data capture must be something along the following lines:

- Craftsman records failure codes and comments on paper copy of the work order.
- Supervisor revises the content of the work order and sends it to the planning department when satisfied that it is sufficient; raises work orders for additional work underway if it is required.
- Planner revises the comments and work order codes and takes the appropriate actions; for example, raising additional work orders, adjusting work order templates, or comparison with the existing information regarding this task on this equipment to see if there are any patterns beginning to develop.
- Maintenance clerk enters the data into the system, ensuring that all of the codes and comments are entered as required to provide solid information for future analysis. You must also include labor hours and perhaps inventory items if your system is not automated in this area.

Although this seems to be a lot of information, some of it may be given via coding. Other parts can be written manually onto the job card itself. Recent advancements, however, have brought us to the point where much information may be entered into a PDA or Palmtop device, then uploaded either in a batch manner or via a wireless computing network, if this exists. This also must be considered if it is required.

Using the wireless technology that is currently available and evolving daily, soon you can have the following scenario (in fact, it is already a reality in some larger installations): A worker requires a part, orders it using a Palmtop device, and the part is delivered within 15 minutes. The worker then finishes the job, updates the Palmtop, and moves on to the next task that is assigned and scheduled for that day.

Another approach to the paperless workplace in maintenance may involve the submission of work requests via bar-coding systems. Coordinated so that they represent the majority of faults that can occur on that equipment, they then are able to simply swipe the code required, which generates a work order complete with parts requirements, etc., from the bank of work order templates. Upon completion of the work order, the technician can simply swipe the codes required and complete the work order. Easy computing in a paperless environment, and it is currently available. As with all of the processes outlined herein, there is a need for exception reporting.

chapter five

Controls and standardization

The most economical way of implementing a CMMS is through a template approach. When looking at a multi-site or multi-company installation, there is often the need to go through similar processes many times. In my early career in the EAM (enterprise asset management) area, I found that there was always the same discussion focusing on the reinvention of processes and codes that I had seen many times before. From this came the idea of the template approach to CMMS implementations.

As a guide, a template provides standardized processes, coding structure, and business rules. Of course, these have been changed in almost every implementation, but they have provided a strong base from which the clients were able to determine what suited them. More often that not, it has been a case of minor adjustments to the overall template.

Apart from the business rules and processes, there are areas in which we can standardize the process of implementation. The benefits from this extend far beyond the implementation project itself. For example, standardization of coding and processes allows the establishment of procedures in each site or company that has the same foundation as a base. This foundation is centered around the needs of the business at the enterprise or corporate level. One of the immediate benefits is in the interchangeability of employees between sites and companies. Because the processes and coding will be the same, there will be little requirement for process retraining.

A second, immediately recognizable benefit is that of corporate-level reporting. With all organizations using similar coding and processes, comparisons can immediately be made between them, as well as having a standardized set of reports to analyze each company or department.

The areas that will be discussed in this chapter range from coding structures to work order template design and usage to standardized forms of creating an asset register.

Work order codes

Work order coding is one of the most vital of the operational codes required. The work order codes will assist in analysis of costs for type of work done, and provide the baseline guide for executing and analyzing a work order.

The work order codes are, in some systems, supplied in two forms: the work order type and the maintenance type codes. Why the difference? Why is it not suitable to record and know only maintenance types that I am working on?

The difference begins at the corporate requirement level, as it always does. There is a need to define the areas where work orders will be used, and the importance of each. In addition, there are severe rules and laws in some countries regarding certain aspects of an organization's performance, such as environmental or safety management.

On the other hand, there is a need to define whether a work order is preventative, predictive, corrective, and so on. There cannot be two types of codes in the same table. A corrective maintenance work order, for example, can be raised for either environmental or safety concerns. As such, there is a many-to-many relationship between the two distinct coding types.

Work order coding definitions also are very useful under the requirements of maintenance work order planning. At any given time, the maintenance planners will be managing a large number of work orders, jobs, and other issues associated with the role of a planner. For them, and anyone else using the system, it is best to have the work organized into easy-to-find groupings that mean something when related to each other. For example, there can be a requirement to review all of the safety work orders that are corrective actions, have a priority of 2, and are in a planned state, or a need to review all of the work orders for a specific project that are capital, and are awaiting materials.

Defining work order types

The division of work into various work order types is driven by the organizational requirements to measure various areas where work is done. This can be governed by regulatory as well as corporate reasons as explained previously.

When defining the types of work orders that are required, there is a need to apply a series of rules. A work order type can most easily be defined as an area of work that has the requirement for work to be carried out in various modes of execution. Basically, the definition of a work order type must be driven by:

- The various types of day-to-day operational requirements of the plant or installation
- Needs to separate or define capital work orders
- Regulatory requirements

However, there is also a set of rules that apply to what a work order type is not:

- Not ambiguous; should be easily understood by all using it
- Not a state of operation, e.g., shutdown or available
- Not a work execution mode, i.e., corrective, preventative, etc., which, as will be explained, are the modes of execution
- Not a maintenance function, e.g., planned/scheduled (the easiest way to render this sort of measure useless, particularly that of maintenance scheduling, is to have it listed as a code somewhere)

As such, the following list of work order types listed can be used as an adaptable guide to most situations and industry types.

General maintenance (GN)

This covers all of the work (regularly scheduled inspections, overhauls, replacements, repairs, and service works) that is required to keep the plant running to set standards. In addition, it covers the emergency works required when the plant is forced to stop in an uncontrolled and unplanned manner.

Capital works (CP)

This covers labor and materials associated with the design, planning, and execution of works associated with the following stages in erecting, dismantling, and improving the plant past its original or current design specifications. This issue has been investigated earlier under the heading of technical change management. It can also be used in devastating situations, e.g., rebuild after a natural disaster.

Following are some examples of capital work orders (dependent on the business rules used):

- Modification within predetermined parameters, e.g., dollar costs, if it forces changes to P&IDs or to the operating procedures of the plant, etc.
- Replacement of equipment or components that have been superseded in some manner or have become too expensive to use for some reason (availability)

Statutory (SY)

This covers all work required to comply with regulatory controls or requirements of a specific plant with regard to electrical systems, pressure vessels, pressure release devices, lifting equipment, etc. SY can be a regularly scheduled task.

Environmental (EN)

EN covers all works required to maintain the plant to the high environmental levels that most operations now aspire to. EN can be a regularly scheduled task.

Safety (SA)

SA includes all works required to maintain a high level of safety. SA can be scheduled tasks and can also include routine safety inspections.

Defining maintenance types

This is a form of executing a task under the various classifications of work defined in the work order types, not a class of work. For example, a task can be executed in breakdown mode but cannot be executed in safety mode. The only exception to this rule is the need for standing work orders.

Maintenance types include:

- Emergency works to restore operations or prevent excessive risk to the safety of people and plant items
- Repairs to plant items, overhauls, workshop repairs, scheduled works, and engineering modifications
- Administrative indicators for overheads

Again, there are some easy-to-follow rules that will allow easier creation of the maintenance type indicators:

- Not a state of operation
- Not ambiguous
- Can be used with more than one class of work

Based on this, the following is generally a good guide to what can be termed as a maintenance type.

Corrective actions (CA)

Any repair work required to return the plant to full operating, safety, or environmental standards, CA can include capital works. For example, repairs after devastating incidents, catastrophic failures, and acts of God.

Preventative maintenance (PM)

PM covers all preventative maintenance tasks for all classes.

Predictive maintenance (PD)

PD is mainly associated with general maintenance.

Standing work orders (ST)

The goal of standing work orders is to reduce the work order load for execution teams, while still allowing effective analysis of data. STs are used across every work class to collect the costs and man-hours on regularly repeated administrative tasks, for example:

- Meetings
- Training
- Consumables, for period, if required

Modifications work (MO)

There are generally a series of criteria describing what constitutes a capital modification within a plant. When an item is below this level, the MO indicator can be used with any work class.

It is important to realize that, depending on the business rules and guides that apply, not all work orders that are modifications will be within the capital class of work order type. For example, there may be a modification to equipment under the safety class of work order types. This may be the extension of a handrail or something similar. In this case, the modification may not comply with any of the requirements of a capital works order and may merely be organized as a modification works order and managed under the technical change process in place.

Workshop repairs (WR)

WR accounts for all workshop repair tasks being undertaken, and includes planned overhauls of equipment and structures. This is a valuable maintenance type which effectively highlights the on-site costs associated with repair works (for comparison).

Breakdowns (BR)

BR covers all breakdowns, for all classes of work. It is an effective indicator for tracking the health of the plant generally. A breakdown must be seen as something that has stopped a piece of equipment and requires fixing or as a failure that will stop a piece of equipment within a relatively short period of time. Whatever the situation, it requires immediate attention.

Work order coding matrix

As can be seen in the work order matrix in Figure 5.1, not all of these codes are interchangeable. In the breakdown example discussed previously, we can clearly see that a breakdown maintenance task would not suit a capital work order type classification because capital works involve construction, major modifications, purchases, or other thought-out, planned, and expensive tasks. A work order with a predictive maintenance type does not belong in this category at all.

However, a modification can occur within all of the work order classes. It is most likely to be associated with capital works, but if there is a modification for an environmental issue that does not comply with the requirements of capital work in the organization, it is allocated in that work order type or class.

Work Order / Maintenance Type Decision Matrix		Corrective Actions CA	Preventative Maintenance PM	Predictive Maintenance PD	Standing Work Order ST	Modifications MO	Workshop Repairs WR	Breakdown BR
General Maintenance	GN	X	X	X	X	X	X	X
Capital	CP	X			X	X		
Statutory	SY	X	X	X	X	X	X	X
Safety	SA	X	X		X	X		X
Environmental	EN	X	X		X	X		X

Figure 5.1 The work definition matrix.

Prioritization

This area is one of the key problems I frequently find in any scheduling or planning system. Very often the prioritization for a plant or company has very little to do with the real priorities of the fault. Prioritization is generally a numbered system with as many as four levels, with descriptions from urgent to noncritical. There is rarely any substance behind these systems in the way of priority codes definition. As such, the definition of each code tends to be left to each individual or, in the case of a meeting, the most senior staff in the room. Either way, the result is poorly prioritized work orders that are not easily used for any part of the scheduling or planning process.

Too many organizations neglect the benefits of a clearly defined prioritization system. When they realize the importance, the focus is invariably at a department or functional level. I have seen organizations where there are as many as three or more prioritization systems, none of which are interrelated. For example, there is one set of priorities for operational maintenance, one for safety tasks, and, at times, even a further one for capital works.

Along with work order classification, failure coding, and integration with business processes, this is one of the key determinants of a maintenance systems future operation. The drawbacks of not clearly defining priorities, or defining them at a departmental level are many:

- Wasted maintenance man-hours on tasks of low relative importance
- Critical tasks being lost in the maintenance backlog
- Dissatisfied operations customers
- Lack of faith in the effectiveness of the maintenance delivery functions

A disciplined method of prioritization will eradicate tasks being done on a whim and allow work to proceed according to its true effect on the overall operations of the plant. It will also allow the maintenance delivery function to be executed in a far more effective manner. For example, while work orders of higher priority will remain those that are scheduled, to achieve the capacity scheduling limits set by the organization, a pool of lesser priority tasks will develop that can be attended to in an unscheduled manner, giving a base for project type works for various craft disciplines.

The last of the benefits that will be achieved by formalizing the prioritization process is the removal of subjectivity and emotion from the process.

System guidelines

The system must cater to the following requirements equally and give a universal method of coding all work orders:

- Site-wide plant equipment priorities, allowing for better direction of resources, which can also apply to various companies if they are of the same type of operation or, at the very least, the same methodology applies
- Operations requirements
- Improvement projects

Accurate prioritization covers three distinct decision-making processes. Although one may be preset, the others will require a degree of judgment and discretion in executing work orders practically.

1. Equipment criticality
2. Effect of task or work to be done
3. Real-world limitations on execution

Original priority of the work orders must be set by the work order originator. This person is the most qualified to answer the questions of equipment criticality and effects of the task/fault noted.

Listings of major equipment and their criticalities will assist in making decisions and lower criticality items or areas will be easier to recognize.

The issue here which does need to be covered by the system chosen is that of real-world limitations. As will be explained later in this chapter, there is a need to adjust priorities to suit the limitations of the operational environment without affecting the overall concept of the prioritization system that is in place.

The coding system described here is an example of how to go about setting a "future proof" prioritization system for work order coding. Although this can appear complex at first, once the required training, documentation, and guidance are in place, it becomes an integrated part of daily business processes.

Equipment Criticality	Description
1	Critical safety protective devices
2	Critical to entire plant operation
3	Critical to continued production of main product
4	Ancillary system to main production process
5	Critical to continued production of secondary products
6	Ancillary system to secondary production process
7	Stand-by unit in a critical system
8	Stand-by unit in a non-critical system
9	Ancillary equipment

Figure 5.2 Equipment criticality listings.

First, there is a need to take into account the equipment criticality that is affected by the fault or work required. This needs to be an attempt to account for the criticality of the equipment as it applies to people and operations.

The criticality table was originally designed for a fairly complex process plant that produced various products (Figure 5.2). As such, it takes into account the needs of ancillary and main product systems.

The other factor to be considered when setting the initial priority of a task or fault is the effect on the plant or on people.

The combination of the criticality and effect of the work can be cross-referenced to give a relative weight of each task in comparison to all other types of works. The colors represent the time frames with which these priority tasks can be allowed to occur.

Although open to debate, the time frames for work orders of varying priorities can generally be grouped into five or less. This allows for both criticality and effect of work, and acts as a cross reference for all work order assignments.

As this is only a tool to reach the time frames represented by the coding, the effects or equipment criticality rating can be changed to represent changes in corporate policy, planning windows, or equipment requirements.

Task Effect	Description
A	Immediate threat to safety of people
B	Immediate threat to the safety of the plant
C	Limiting operations ability to meet primary targets
D	Limiting operations ability to meet secondary targets
E	Hazardous situation, people or machinery, not immediate
F	Will effect operations after time
G	Improve the efficiency of the production process
H	Restoration of the plant technical integrity
I	General improvement to further operability, safety or maintainability goals

Figure 5.3 Task effect listings.

Real-world priorities

In executing this work, there will always be difficulties associated with labor shortages, plant availability, and materials lead times. To balance these items against the priorities of the plant as represented by the originator, the planner requires a system of coding work orders to determine how they will be completed practically. The degree of flexibility must be built into the coding system. This adds a third dimension to the prioritization process and allows for the consideration of real-world limitations on workflows, including rapid changes to overall operations priorities. In order to carry out this facet of the prioritization, without affecting the original priority given to the work, an additional field will need to be found or provided.

Although there are many recommended approaches to prioritization, I would recommend a matrix-style approach that is suited to organizations where the maintenance function may cover various plants or ranges of equipment, and can be adapted easily if the organization decides to restructure itself (as happens frequently).

No matter which approach is used, the maintenance department cannot continue to carry out its function with any degree of certainty without it.

	Effects of Task								
	A	B	C	D	E	F	G	H	I
1	1A	1B	1C	1D	1E	1F	1G	1H	1I
2	2A	2B	2C	2D	2E	2F	2G	2H	2I
3	3A	3B	3C	3D	3E	3F	3G	3H	3I
4	4A	4B	4C	4D	4E	4F	4G	4H	4I
5	5A	5B	5C	5D	5E	5F	5G	5H	5I
6	6A	6B	6C	6D	6E	6F	6G	6H	6I
7	7A	7B	7C	7D	7E	7F	7G	7H	7I
8	8A	8B	8C	8D	8E	8F	8G	8H	8I
9	9A	9B	9C	9D	9E	9F	9G	9H	9I

Figure 5.4

Time Frame	Color	User Priority
24hrs		1
48hrs		2
72hrs		3
3 week		4
3 weeks +		5

Figure 5.5

Equipment register standards

For some reason, there is always a lot of discussion about how the equipment register should be constructed and formed. I have seen various approaches to this, even one in which numbers were drawn to group equipment (which actually was not bad, apart from some "search" issues).

An asset register must be created in a focused manner and based on the functionality of the equipment rather than its geographic location. When attributing costs through a work order to a current carrying cable that is part of the electricity supply network, the costs need to be attributed to the costs required for supply of electricity, not for the geographical area. The same applies to almost all equipment. For the purposes of the equipment or asset register, there is no need for grouping by geographic location. However, there is a need for future searches to list exactly where in a plant the equipment is located.

Work order templates

Today, most CMMSs will contain work order template functions of some kind and these will go by various names. However, as the name here implies, these are templates for work orders that will need to be created at some stage or another. By being able to create a work order template, a whole series of opportunities opens up. These will become the experience library, allowing us to pass our learning on to future employees and to learn from past experiences.

So what exactly does a work order template do? This function of modern day CMMS is useful for creating a standard of how to do work. The work may be corrective, preventative, predictive, or even some modifications that you are likely to repeat. And this is the clue — work that you will need to repeat.

A template can be linked with a maintenance routine or used to create a work order when there is a need to do so. By linking it to a maintenance routine, you should be in a situation that leads to a one-to-many relationship, i.e., one template work order to many maintenance routines. For example, a plant may have 50 or more of a particular type of pump. A work order template can be linked with the routine services of all 50 of these pumps.

The benefit of the one-to-many relationship is that when a change is required, it only needs to be made on the one template, and this should reflect on all 50 of the connected routine services.

As well as being the repository of experience for your maintenance plant, a template is also a tool for cutting down on the amount of planning work required for each task. Within each template, you should have all of the relevant information required to complete a task, and it must align with the corporate definition of what constitutes a planned work order.

The template can also be used to contain all of the planning information that is required for any corrective task. Here we start to get into an interesting area. The definition we used of a planned work order was one that was ready to be executed. In the event of a breakdown of some sort, you have a template available to create a "planned breakdown work order." That is a reactionary task that was "unplanned" but is "planned" and ready for execution. In the section on planned/scheduled ratios, I eluded to this point. The point was, instead of an emergency or breakdown work order being unplanned and unscheduled, it is in fact planned and scheduled.

Another function of the work order template is to provide leverage over the planner-to-craft-employees ratios. There is an often-quoted figure that a maintenance planner should be planning for no more than between 15 and 20 craftsmen. With the effective use of maintenance work order templates, you will be looking at the possibility of extending that to between 35 and 40. The reason is simple: with the advent of more and more templates, of higher and higher quality, the amount of time spent by the maintenance planner on the areas of work order planning is considerably reduced.

The contents of a work order template can include some or all of the following items, depending on what is required from the planning function.

The estimated resource hours and types

You need to keep an accurate record of the resource hours that you will require, as well as the resource types. However, the work of estimating resources does not stop with the template creation. The goal of any team doing repeating work is to do so in a manner that it is ever-reducing in time and resources while maintaining the required quality and safety levels.

The estimated materials

Although quite distinct from MRP (material resource planning)-style inventory managements, you still need to adequately manage your anticipated demand for various materials. Within a CMMS, this information will go a long way to allowing you to optimize your inventory levels to a point where you are holding the least amount possible and still maintaining enough stock to provide service levels between 85 and 95%.

I will review this in greater detail when I speak of the maintenance store. However, the service level of an item is the level at which you are able to supply the maintenance requirements 85 to 95% of the time that it is required.

There have been books written claiming that anything below a service level of 95 to 98% is unacceptable. This is not true and it will depend on a multitude of issues which I will discuss.

The estimated costs

Materials such as parts and consumables, as well as human resources and any other costs such as equipment rental or contractor costs, also need to be estimated. When you are revising your backlog, you need to know the estimated costs of your backlog work, in addition to knowing what are the planned resources and materials over the next scheduling period or entire backlog content. This will enable you to look forward as to whether your budget is enough or even if it is too much.

The human resources estimates also need to be under regular review. Your cost estimates and resource estimates will help to determine your overall effectiveness as a maintenance department.

The procedure

When I discussed job packaging and planning, I touched on the need to attach documents to the work orders, either as a part of the text of a template or as an attached document. The advantage of doing so is that you do not have to do so in the future. There is no need to continually reassess what documentation is required every time you change planners or procedures in some form or another. The documents will always be attached, and when you create work orders via your routines, you will automatically have the link to the document.

Again, by linking documents you are able to change the information in one place and have that change reflected throughout the entire work order range that uses that document. Similarly, when you change the procedure in the text fields of a work order, the text that appears on every work order that is created using that template is changed.

This functionality can be used to include "tips" or updates to the procedures on how to do the work. For example, when a new way of doing the task is uncovered that saves time, eliminates frequent maintenance rework, or is a safer way of doing the task, it should be included in the procedure in some form. This is the manner in which your work order templates become the repository of experience and one of the centers of your continuous improvement programs.

Safety information

The need for a strong focus in the execution of maintenance work is now a known fact and not something that is necessary to reinforce here. So, the template must to have relevant safety information, either as a linked document or as a part of the text of the template. This may be a MSDS (Material

Safety Data Sheet), a risks analysis, or a list of required, protective safety equipment; whatever way, there is a strong need for its inclusion.

Other information that can be determined as safety information is the types of permit required, or even a blank permit as an attached document. As a means of increased safety management, there has been a shift toward lock-out safety tagging isolation systems over the past ten years. Although useful in most industry sectors, there are those that require it more than others. For example, in the electrical generation, transmission, and distribution networks of the utilities sector, regulated and enforced safety procedures are an essential part of keeping workers alive.

Specialized tools and equipment

Although often included in well-written procedures, this area does need to be covered so that we may be able to include the information needed for ease of scheduling and ease of execution. Often in heavy industry, there are requirements outside of the usual that can affect the planning and scheduling of tasks. Examples of this are large cranes, scaffolding structures, or scissor lift devices — items that do not normally form a part of regular maintenance tool stores.

Standardized text

One of the often-overlooked benefits associated with work order templates is the ability to standardize the information that appears later on work orders. With this function, searches can easily be done on the fault description for a specific piece of equipment as a part of your analysis procedures. Although this can be covered by the many work order codes that are available, it can assist to also have this option available.

Another area that comes under standardized text is the standardization of work order codes on work order templates. Each template should have, as a minimum, the type of maintenance task, the type or class of work order, and any other relevant codes that can be used to identify and later analyze the work. This can go as far as the failure coding which will be discussed later.

As can be seen, the concept of the work order template for maintenance purposes is extremely useful and should form a part of any CMMS installation. This function will provide the benefits of ease of pre-planning and easy use of this information to state what your upcoming resource requirements are, with little effort required. The planning staff can then begin to dedicate more of its time to other facets of the maintenance planning role, as required.

However, of more value than any other potential benefit is the benefit of having a well-maintained repository of experience. This can become the basis of continuous improvement efforts and will be vital in surviving changes in the workforce such as transfers, leave arrangements, promotions, and resignations. You are able to come one step closer to the situation where the system is of greater value than any of the people involved in it.

Failure and completion codes

The area of failure analysis is one of the best reasons to purchase and implement a CMMS. The ability to capture data by integrating the system into the daily activities of all involved in the maintenance effort is where you can easily reduce many costs associated with low availability or low reliability. Special attention must be given to the selection of your CMMS in order to extract this form of information.

First, it needs to be understood exactly what it is that you are after. Here you need to look at the systems that are currently in place and proven in this area, principally the system of RCM.

Failure coding tends to be associated within major EAM systems and the closing of work orders. Generally, these are codes that can be defined by the clients or end users of a system. Some commonly used codes are the following:

- What was the cause of the failure?
- What was the fault noted?
- What work was done?
- What is required to avoid this in the future?
- Is a root cause analysis required?

It should be mentioned here that there are no codes for "What was the root cause of the failure?" because the person doing the work is not likely to know. If you were to look at all of the root causes that could be listed for all of the possible failures of all of the equipment that you have, you would end up with a list that is far too long and extremely unusable. The project that attempts to determine this would be extremely large and not a productive use of time. It is far more useful to commission root cause failure analysis projects and teams to attack the problems that are perceived as issues. The codes here will serve as a guide as to what does or does not require a RCFA project. But it will most likely come from other sources including anecdotal and cost analysis reports (among others). There is also no focus on the consequences of the failure as that can also create a great multitude of information. The basic reason for this form of information is to use it as a guide to the problems you have on specific equipment and how frequently.

Some of the CMMS style systems in today's marketplace are very much focused on having a "smart" listing, which means that the system recognizes the type of equipment that you are entering codes for and shows only the codes that apply to that type of equipment.

In short, you need to develop, and afterward filter and produce reports on specific failure causes for a piece of equipment, as well as a list of specific work that can be done in order to correct these failures. From this, your failure reporting will focus not only on the frequency of particular failures but also on the steps taken to repair or overcome these failures. From here, you are in a good position to commence the failure analysis and embark on the RCFA methodology that you have adopted.

In some cases, this functionality does not apply. Therefore, every time you enter a failure code for a work order on a piece of equipment, you are confronted with a long list of failure codes. The problem here is human nature. When presented with a long list where there is a need to filter through the information to find what you are looking for, the preference is not to do it. In addition, the chance of error is much greater in this fashion. When a pump fails, you want only to see the codes that apply for that pump, and you have no interest in seeing the codes that apply to a high voltage transformer in the same plant.

Another area where you need coding is in the fault noted. On the whole, the better option is to apply codes learned or developed during an RCM-style equipment analysis.

chapter six

The maintenance store

General

The interrelation and synergy between maintenance and its inventory management partners are the strongest links that exist in any organization dependent on maintenance of its physical asset base. Along with the functions of maintenance, there has been tremendous evolution and advances in the area of materials and inventory controls.

Traditional stores management characteristics

- High inventory levels
- High levels of stock outs
- Many unofficial stores under user control
- No measurement techniques
- No management interest
- All items given equal attention (the famous 95% for everything regime)

Advanced Methodologies

- Different management for different items
- Different levels for different items
- Inventory levels optimized
- Controlled by inventory management professionals
- Justification for stocking catalogued items required
- Few or no "unofficial" stores
- Inventory performance importance
- Use of simulation tools
- Monitoring of key performance indicators (KPIs)
- Use of E-commerce methods

If the maintenance store does not have the items you require, that impacts directly on the ability of maintenance to fulfill its role in the organization. Even worse, if the details in the CMMS are incorrect, the maintenance of equipment may be delayed by unexpected waiting for components or parts.

On the other side of this equation, it is easy for a maintenance store to become bloated with unnecessary items or items with low cycle rates. In this situation, you are wasting capital by having unused items for extended periods; as a result, you are incurring maintenance costs by having the material in the maintenance store.

The inventory management of items for maintenance consumption must be managed in a very special "just in case" style of arrangement. On the other hand, there are three basic reasons that the JIT (just in time) methodologies that are gaining popularity do not work nor apply here:

1. Maintenance of equipment often takes in the purchase of items that are specialized or have very few providers in the world today.
2. Many of the large plants and industrial complexes exist in isolated areas where lead time is a major factor.
3. As will be explained later in this chapter, JIT-style management does not take into account equipment criticality, i.e., when an item breaks, and this can happen in random failure patterns, there is no understanding that you need this item NOW!!

Conversely, the levels of inventory to be managed are very dependent on the performance of your equipment. Equipment with higher levels of availability and reliability require fewer urgent parts dispatches. They also require fewer items of long lead times, due to a higher level of reliability engineering focus. As such, the inventory levels in this respect can easily be maintained at very low levels in an easy manner, thus saving you invested capital.

There are other side effects of low equipment reliability. The fact that you never really know what is required from day to day leads you to a situation where inventory departments have an even less idea of what you require.

There are two possible outcomes. First, you begin to stock a great number of items and at higher-than-required levels. Second, maintenance loses confidence in the stores system and begins to hoard items. This, in turn, generates uncontrolled stores, and the system breaks down. The uncontrolled stores become the first port of call of maintenance people, and they begin to manage it themselves in a manner that is generally not economical. In an extreme case I have seen them begin to manage their own repair items, further distorting the stores' inventory levels and management processes.

In short, the function of inventory management is to provide strong customer service while optimizing the use of capital dedicated to its management. There is no CMMS in the world today that will adequately do both of these functions. As often mentioned throughout this book, it is a combination of sound management practices and strong software functionality.

The management of inventories in a modern-day CMMS must be done in a very quick and easy manner. Maintenance must be able to see a variety of information quickly and the inventory department must quickly react to maintenance requests. From a maintenance perspective, they first need to review the items and the respective levels that are available within the maintenance store. From that point, there is a large range of information and functions that the maintenance department will need to have available to it, including:

- Levels of stock on hand at that moment
- Number of items ordered and when they will arrive
- Reservations of items to other work orders
- Lead times of components
- All relevant stock item details
- Possible substitute items
- Where the item is (this can be useful in the case of a multi-warehouse or multi-site organization, when an item that is not available locally maybe be available at other locales)
- For ease of picking, it may also be required to determine the aisle or inventory bin the item is in
- Full range of search criteria:
  - Text descriptions
  - Stock codes or part numbers
  - Manufacturers' codes
  - Commonly used names for the items (it is more than likely that the people on the shop floor have a different term for an item than what the manufacturer calls it)
  - The equipment it is used on

From the inventory management side of things, there is a range of other information that must be available to them:

- Item classification
- Service level requirements
- Reorder points and quantities
- Other details that determine the item classification, service level requirements, and reorder points and quantities, such as details that are best grouped under the heading of ABC analysis
- Analytical reporting to readjust levels of stock.
- The ability to do stores recounts to ensure the accuracy of the system data

There are two distinct types of warehouse localization methods: (1) centralized and (2) localized stores. Within each of these stores, there are also three distinct forms to manage access: (1) open and (2) restricted, and (3) a combination of the two.

The centralized option of stores management is often the better way to manage these items. With a centralized system, there is less likelihood of inventory levels expanding much past economically required levels, and there are substantially less administrative resources required to manage the system. However, the tradeoff is in the ease of access by maintenance people. In a centralized fashion, the maintenance people will have to travel more to receive the parts required. By implementing a delivery system, this may be circumvented, but it will still exist for urgent items. If there is an urgent need for operations to have direct maintenance access to items, this may not be best suited to your organization. However, I would question that need as possibly being related to uncertain and poor equipment performance more than any other issue.

The issue of stores access is an interesting one. Many stores organizations may have attempted a change to open stores access, allowing the maintenance people to get what they require and book it out at the same time. The focus here is reduction of administrative staffing levels. I have yet to see this work in practice. The most common result of this style of stores access management is missing items and unreliable stores information, which then translates into lack of maintenance confidence, the creation of uncontrolled stores, and general lack of inventory control.

The organizational requirement of your system is to focus on the authority levels of each person with access to create warehouse requisitions and purchase orders. A function that will assist you greatly is where you are able to have limits in dollar values associated with each position. This also must be arranged in such a manner that a requisition or purchase order that is outside of the range of a particular person will be directed to his superior for authorization or directly to someone whom your organization states is responsible for such things.

As discussed previously, there is a need to classify items in the maintenance store. This will determine how they are managed in the corporation, and may also trigger other functions in the CMMS. Basically, there are a number of different classifications that you can apply to your inventory items, whether they are critical spares, repairable spares, insurance spares, items to be traced, and so on. The two main items here for your maintenance considerations are repairable items and those items that are to be traced.

An item that can be repaired is an item that, due to initial costs or lead times, is more economical for you to repair than to replace. In this case, you need a system within the CMMS to handle this. Repairable item management is a very economical form of asset management in which you can save a great deal of money if it is implemented in a correct manner.

The system must be associated with the automatic ability to know when a repairable item has been booked out of the store, so that the process of rebuilding the item it is to replace can commence. From there, you need to have a record of who is the most reliable supplier of these repair services.

This applies to the CMMS's ability to report on the performance of your suppliers. For instance, you need to know:

- The average time to repair the item per supplier of the repair service
- The number of times the item has been returned after agreed deadlines
- The amount of price variation in repairs

This information can be the beginning of a supplier relationship management function and the item can be sent to where it is best suited for repair. The lead time of this repair also must be entered in your system, as well as some form of notification when it has returned. This functionality is important for you whether the item is to be repaired on- or off-site.

You need to associate the costs of on-site repairs to a "workshop repair" maintenance-type work order for future analysis. With off-site repairs, the cost is easily traced. However, with on-site repairs the costs can easily become obscured by the work order being buried in a sea of like-coded orders. By associating on-site repairs to a workshop repair work order, you can identify the work order easily.

The other item of special importance is that of traceable items or items that you want to follow through the course of their useful life. This may be items that are either very expensive or are very critical to the operation. Any item will have various points in its life cycle that are important for you to know. For example:

- Purchased
- Installed to machinery
- Removed for repair
- Repaired off-site or repaired on-site
- Returned to the maintenance store
- Discarded after several passes through this cycle

Your system must be able to capture these points and provide you with information on the overall useful life of these components, as compared to the life you bought them for, as well as all of the costs associated with this component during the fullness of its life span for future budgeting and analysis purposes.

Service levels and inventory policies

Service levels of components or spare parts are basically what percentage of time this item is or should be available when the maintenance department requires it. The service level of any item goes a long way to determining the inventory management policies and to determining how you will manage this item.

The service level is set using a number of criteria. First, there is the criticality to maintenance. Here the system needs to record this point. Second, the price of the component, both to purchase and to maintain in the store, and third, the availability of the item.

As such, there will be different service level requirements on each of these items, depending on the combination of the three factors mentioned previously. It is often stated that the service levels of items must be 95%. This is not correct, and will lead to inventory levels being managed in a less than optimal manner.

In large stores systems, there is a tremendous amount of time that must be dedicated to the management of inventory levels. From the outset you need to classify your items into (1) what are you going to actively manage and (2) how these are going to be managed. The average time for a stores controller may not be distributed evenly. In fact, once you calculate all of their availability divided by the number of items stored, you may find that you are talking about a number of minutes per item. For this reason, you need to classify what item gets what level of attention.

The most frequently used and effective method is that of the ABC analysis, which is directed at determining the attention span per item and can be described as:

A – Much attention
B – Normal attention
C – Less attention

How, then, do you go about determining what an "A" level classification is and how it is treated? This goes to your stores analysis and policy procedures. Although each company may differ somewhat, the criteria for slotting an item into any of these areas depends greatly on cost, availability, and criticality to the operation. Of these three considerations, it is maintenance criticality that is the hardest to work out and often becomes an area of contention. Criticality should, as with anything else, be determined in the simplest manner possible, something along the lines of the following:

A – Will stop production
B – Can be delayed without effect
C – No effect on production

The other areas can also be qualified in a similar manner.

Availability of the item

A – Greater than one week
B – Greater than one day
C – Within one to two hours

Cost of the item

A – Greater than $10,000
B – Greater than $5000
C – Less than $5000

So you see that the ABC analysis is really quite deep in what it represents; therefore, you need something within the system to manage and something that you use to set service levels. For example, a part with the ABC rating of BCC may need a service level rating of less than 60%. Why not? If it can be had within one or two hours, the item will not immediately stop production and it is of low cost. Why should you dedicate a lot of time to its management and stock levels? In fact, it would be worth evaluating whether it was required to maintain an item in stock with such a rating.

However, an item with a rating of AAA is a different story. You need to manage these items manually for a start. Many systems have automatic reordering functions, but these would not be adequate for such a high criticality, low availability, and high-cost item. As such, you need to be monitoring it frequently.

There is also a range of other factors that need to be applied and included in any inventory management system, all of which will go to defining their classifications and how they will be managed in the future. This includes the store levels and the levels of reordering. These factors include the ability to find substitutions and the items' annual usage rates.

Optimization levers

Optimization of maintenance inventory must be handled in two distinct forms: (1) by being able to draw reports for analysis, and (2) by being able to apply simulated scenarios to the item.

In terms of reporting, there are a few basic reports that you require from the system. For one thing, you require the report on slow-moving items over the last year. This gives you a starting point to determine the items you can revise in conjunction with maintenance in order to determine the possibility of obsolescence. In addition, you need a report of potentially obsolete items. This report must be focused at a period of greater than a year. From this point, you can then determine if the item is still required according to the ABC analysis method.

Other reports that you need are:

- New items added for period and their usage
- Number of stockouts and their effects
- Number of times an item has not complied with its service levels

With this as a base, the CMMS will be able to manage the inventory requirements of your installation in a relatively optimal manner. This topic is one of great depth and detail. Here you are only looking at baseline requirements as they apply to your CMMS implementation template.

In terms of simulation, you need to see the sawtooth graphic that represents an item's inventory levels over a given period. Then, with manipulation of reorder points and reorder quantities, you need to see the possible economic effects of different levels. This is one of the greater tools available on the inventory management side of CMMSs today and should be considered in any purchase.

The Internet

In recent times, the Internet explosion has affected many industries and sectors of the economy. One of the most affected areas has been inventory management and purchasing. There are now a myriad of sites on the Internet that are E-marketplaces — basically collections of product and service providers and their prospective customers. One in particular was recently created by some of the major mining houses and their equally large suppliers.

This greatly facilitates E-commerce applications or buying over the Internet. There are also sites where companies can list their obsolete or used parts for sale. What does that mean for you? Basically, it means that you can log on to suppliers catalogues to find the items you are looking for, and you can find them at a location closer to you and at the price that is generally best for you. More than anything, this simplifies the purchasing process. One of the key elements in simplifying purchasing is to reduce the number of vendors in order to get higher quality service and products from the remaining vendors.

One of the other very interesting and timesaving developments is in the area of graphic parts books associated with some of the major CMMS makers. These parts books are a way of graphically depicting the components of a piece of equipment to be able to easily select the items required for ordering.

One of the benefits is that they are able to link with vendor information on the Internet. This means that the catalogue of a large haul truck or a pump can possibly be found and downloaded or updated directly to the CMMS. The time saved in cataloging of items is immense, and it is one of the quickest ways to maintain the stores' inventory. In addition, procedure manuals and parts listings can be used within this medium.

chapter seven

Reporting and KPI development

With the processes, procedures, and main areas of your CMMS requirement now completed, your attention needs to turn to how you measure this, and what form of reporting is needed for you to use the system. There have been several books written on the subject of KPIs (key performance indicators) or metrics. While they are all very good, they assume that the information that you require to measure exists in the system.

Based on the methods and approaches outlined in the previous chapters, you can be very sure that the information does exist. You can proceed to the definition of the KPIs and reports that you will need. There are three main areas of reporting and metrics that you need to consider:

1. *KPIs and measures:* Reports focused on the measurement of key performance indicators to identify problem areas that need your attention
2. *Functional reports:* Reports that you require on a daily basis in order to carry out the work in accordance with the processes you have defined
3. *Exception reports:* Reports that you require in order to measure compliance with the business rules that you have set (discussed in previous chapters)

Although most CMMSs are delivered with a set of standard reports, there is a need to create your own reports quickly if the system reports do not accommodate your requirements within their standard scope.

KPI development

KPI development is of key importance to the final outcomes that you get from a CMMS, although more often than not, there will be a widely used KPI structure in place. It is of vital importance that you revise this to better define your CMMS requirements, as well as redefine your corporate focus in this process.

The focus should be threefold: (1) on the definition of the KPIs that you will be using to better measure equipment performance, (2) on the KPIs that you will be using to define financial performance, and (3) on the performance of maintenance processes generally. There are various sub-headings of each of these which will also be of use.

The overall business requirements first need to be defined. In the case of a mine site, you can state that the production plan for the year is based on a given availability, or, in the case of a process plant, that throughput for the year also will be based on predetermined availability statistics. As with everything in the model that you are developing, there is a need to determine the business requirements first.

Following are a series of maintenance indicators that are of use in any situation. However, along with the indicators themselves, there needs to be a guide for how they are to be used and applied. The principal reason for the use of KPIs is not to produce pretty graphs or to tell you when you are doing well. Rather, it is to indicate when you are not doing well, and provide a point of reference that you can use in finding the fault in your processes or strategies in order to correct it.

Equipment performance

Some of these will be recognized as standard measures, while others will be new to you. However, most are proven measures in today's industrial environments.

Availability

First, there is a need to define the overall downtime or availability hierarchy as it exists in your operations, i.e., what each of these measures is focused on, and how you classify different periods of time, depending on what they are used for in your organization.

Figure 7.1 gives a high-level representation of this and attempts to describe the process by which you can classify these periods. The table represents three levels of downtime analysis only; however, it can be used to apply to many more levels of operations time. This will depend on the level of detail that the organization wants and also on the level of explanation that is required. As always, there is a common rule that applies. From this there is the ability to define various KPIs and graphics of how the organization is performing. I will limit further examples on this theme to a fleet of haul trucks or transport trucks in order to provide clear explanations. Reference will be made to other areas; however, the main focus will be on this example.

Level 1	Total Time (TT)					
Level 2	Operation Time (TT)			Maintenance Time (MT)		
Level 3	Productive Utilized (PU)	Unproductive Utilized (UU)	Idle Time (IT)	Breakdowns (BR)	Unscheduled Maintenance (UM)	Scheduled Maintenance (SM)

Figure 7.1 High level availability.

If this hierarchy or one similar to it is to be used by your maintenance and operational measures, then there is a need to define this requirement in some manner when compiling the CMMS template. This will become clearer when you actually discuss the template itself. In addition, you will need to decide if you need various availability hierarchies to be determined, e.g., different hierarchies for different pieces of equipment or, through some creative coding, using only one system.

Total time (TT)

Total time is the sum of all time required by the operational departments for use of the machinery. In the case of a 24-hour operation, there is a need to focus this on the total calendar time available. In the case of an operation that works only 8 or 16 hours per day, there is a need to focus this on that period only.

Operations time (OT)

This is all of the time used by operations within their various functions. For example, time for operations may be divided into three main areas:

1. Productive utilization
2. Unproductive utilization
3. Idle time

Depending on the focus of your operations, there may be a need to determine a fourth area, operations-caused downtime. Returning to the example of a hauling fleet at a mine site, there may be a need to attribute all tire failures to operations. The major factor affecting a downtime period such as this is that of the road condition, a factor over which maintenance may not have control. A second criteria may be that of operational errors, items where the operation of equipment is the primary factor in its failure. Although this can be very useful, it can also be very contentious, and must be applied in situations where the corporation is willing to use the findings to improve operations procedures. It should not be defined or used in an organization where it would be used as a trigger to dismiss an employee or place the blame somewhere. Such practices lead to inaccurate recording of results and can also lead to lower employee morale.

In addition, if the creation of such indicators will cause problems and conflict between the two departments, it would be wise to focus instead on creating a synergistic and focused environment within the organization prior to implementing an operational error code.

Productive utilization (PU)

Productive utilization codes cover the main areas of operation where the equipment is used to fulfill its principal functions. Here, you need to clarify that it does not matter if the equipment is used to its capacity, only that it is being used to fulfill its primary function. In the case of a hauling truck, its functions including waiting for a load, loading, hauling, and dumping. As a matter of fact, the four functions described here would be ideal as subclasses of productive utilization.

As can be seen, the construction of an availability hierarchy is a useful tool for all aspects of an operation, not only measurement of maintenance performance. For example, a report on the percentages of productive utilization time may show that there is a disproportionate amount of time waiting to be loaded by the hauling units in the fleet. This may lead to the adoption of different loading procedures and possibly more loading units.

You may see the possibility of taking this to yet another level of development. For example, under travel time (TR), you can calculate travel to the loading unit and travel to the dump site, vital factors which you need to consider when planning this hierarchy.

Unproductive utilization (UU)

Unproductive utilization refers to time where you are not using the equipment for its primary functions. For example, it can be time where the operator is taking a break or attending a toolbox meeting, or when the equipment is undergoing a pre-shift inspection. Although there is no doubt that these periods are productive and useful from a corporate point of view, they are not productive uses of the equipment, and this distinction needs to be made and communicated.

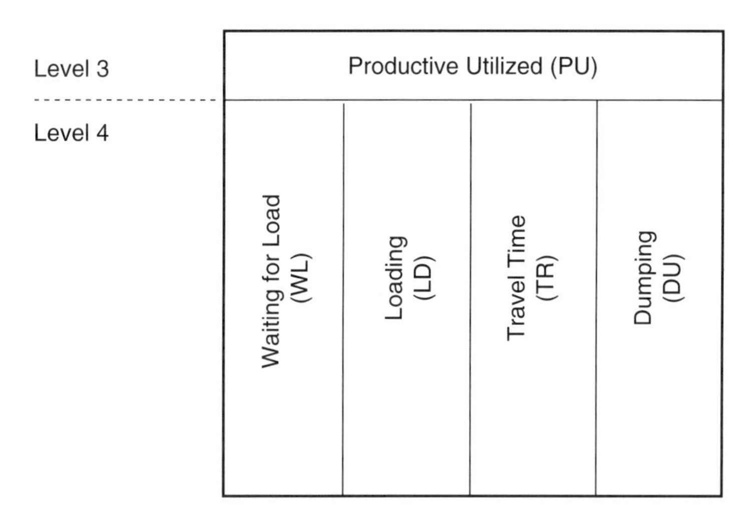

Figure 7.2 Productive utilized breakdown.

Idle time

This measurement is critical to the optimization of your operations in the future. Idle time describes all of the time that the equipment was available and not used. There can be a myriad of reasons, however it is vitally important that you measure and record this because it can affect future planning and efficiency targets. Referring again to the model of the hauling truck fleet, idle time can apply to

- Time unused due to production requirements
- Time unused due to meetings or other organizational factors

Maintenance time (MT)

This is all of the time that a piece of equipment is not available for reasons associated with the maintenance function of a the company or fleet. This may be time for scheduled or unscheduled maintenance or time for breakdowns, as defined by your corporate guidelines.

Breakdown (BR)

BR is the total time the equipment cannot be used due to it having a fault of some kind. There is a need to define here that a breakdown is a corrective maintenance action. However, it is one that has stopped you from operating and needs to be fixed prior to going forward. Therefore, when you have a breakdown that is serious and it takes the equipment away from operations for periods of several days or weeks, there is no time, from the point of view of availability recording, when this will become anything other than a breakdown. If you plan the work, this will factor heavily on your work order management KPIs, but it will not matter to your availability calculations. A breakdown is a breakdown and in order to realize great improvements, you need to raise the bar.

Depending on what equipment you are operating, you may be able to define a substructure under the area of breakdown. A warning here: please do not try to define every possible area where you can have an equipment breakdown. Within most of the major systems on the market today, there is an ability to link the recording of codes to a work order that was raised to do the work. As such, the bulk of the definition and detail will be in the work order, while the classification will be in the coding.

Scheduled maintenance

Again, you need to define definite "horizons" or limits within which you consider it important to comply with the scheduled maintenance definition. I suggest that if the task existed on the weekly plan prior to your agreement and to starting work on it, then the maintenance is scheduled to be done. There is no doubt that you can schedule work within a one- or two-day horizon. However, your goal should initially be to achieve a scheduling horizon of one week, and you need to mark against that. As you get better at it, your indicators will reflect that and you then need to contemplate pushing your horizon to two or more weeks. Scheduled maintenance includes all of the maintenance work that you have included on your weekly schedule and falls under the planning and scheduling category of planned/scheduled.

It does not matter if the work is corrective, preventative, predictive, modifications or workshop repairs. If it was scheduled at the beginning of the period, whatever that may be, then its outcome is to be recorded as scheduled maintenance.

Again, you find yourself in need of further levels to the original hierarchy table. You are in need of a definition principally of whether it is electrical or mechanical. These are the two main discipline areas which can be further broken up as shown in Figure 7.3.

As with all items in this book, this is very definitely an example. There are many ways to define downtime codes and many ways to apply them. This is merely the better of the procedures that I have come across in the areas that I have consulted.

Unscheduled maintenance

This is a means of measuring the time dedicated to opportune maintenance or maintenance that you are scheduling within your one-week horizon. Opportune maintenance and maintenance that you are scheduling can also be two sub-headings of the next levels of the hierarchy. Whatever structure

Level 3	Scheduled Maintenance (SM)					
Level 4	Electrical (EL)		Mechanical (MC)			
Level 5	Electrical Controls (EN)	Instruments (IT)	Pneumatic Items (PN)	Hydraulic Items (HD)	Mechanical Items (MI)	Engine (IT)

Figure 7.3 Scheduled maintenance breakdown.

you use, it needs to be seen that opportune maintenance is not detrimental to overall operations.

There are a multitude of reasons that opportune periods are often created in various processes. Perhaps you do not have access to product to mine, or you do not have the primary materials to continue with the manufacturing process, or you do not have demand for the product. Opportune maintenance can even be caused by the breakdown of one piece of equipment, leaving an associated equipment available for maintenance. In any case, the result is a window of opportunity in which you have access to the machinery or an opportune period for maintenance purposes.

The other form of unscheduled maintenance that you can perform under this heading is that of unscheduled corrective actions. You may recall that you are operating under a weekly time horizon, and, as stated earlier, anything that is scheduled within that week, although technically a scheduled task, needs to be recorded as unscheduled maintenance. Urgent items arising from a routine inspection or other such things would fall into this category.

There is a need to define the electrical and mechanical sections of this part of the hierarchy and their relevant sub-headings or categories, depending on where in the system you place them.

The formula

Now that you have defined a maintenance downtime hierarchy, you need to consider the types of availability formulas that are available.

Standard availability

Total time (TT) – maintenance time (MT)/total time (TT)

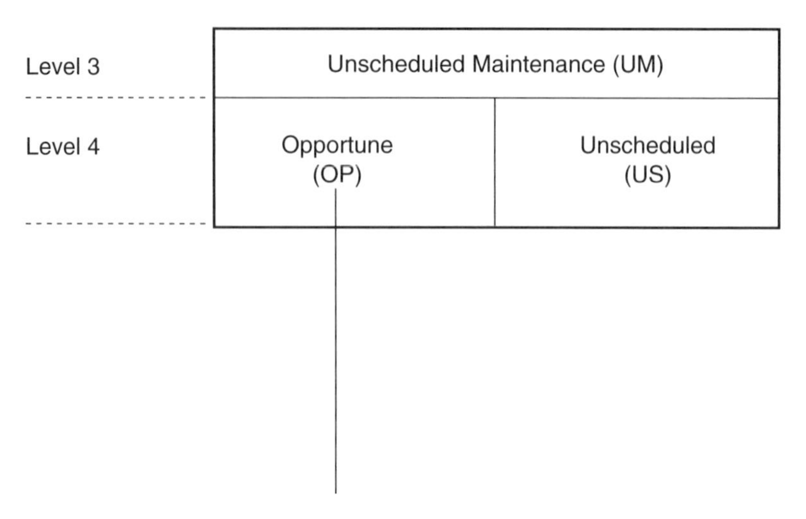

Figure 7.4 Unscheduled maintenance breakdown.

The goal of this formula is to determine the availability of the equipment to operations during a given period. Depending on how far your organization wants to go into the creation of a downtime or availability hierarchy, there may be cause to further define the formula:

Total time (TT) – (SM + BR + US)/total time (TT)

However your organization decides to define this, it is important to do so with the following in mind. It is only the time when the machinery was not available due to maintenance work, whether predetermined or otherwise. It is not a measure that should include the time when the equipment is given to maintenance, such as opportune periods.

Mechanical availability

Total time (TT) – total mechanical maintenance downtime/total time (TT)

The goal here is to identify the amount of time that the equipment was available from a purely mechanical point of view. Another way of describing this is the performance in a mechanical sense of the equipment.

Electrical availability

Total time (TT) – total electrical maintenance downtime/total time (TT)

Utilization

Productive utilization + unproductive utilization/
total time (TT) – maintenance time (MT)

The goal of the utilization measure is to determine what percentage of the time available you actually took advantage of. Again, using further levels in your hierarchy, you can further define what each piece of equipment was doing, but the focus here is only on the utilization measure.

Effective utilization

Productive utilization/total time (TT) – maintenance time (MT)

The goal of the effective utilization measure is to determine what percentage of the time available you took effective advantage of. This measure, more than that of utilization alone, can show what sorts of problems you are having on a regular basis so that you can encourage your operations departments to better utilize their equipment. As always, you need to bear in mind the following: the equipment belongs to operations, but the reliability of it belongs to maintenance.

As such, again referring to the haul truck example, it is easy to see that if you raise availability and effective utilization, you are in a position where you may be able to lower committed resources (in this case, haul truck units). The savings here are massive and well worth the time taken to measure and analyze the numbers. For example, you can realize saving in the areas of:

- Operational costs
- Maintenance costs
- Inventory costs

Other measures

There are other more graphic means of producing these measures than merely in the forms stated previously. Figure 7.5 is a visual representation of the time used in the various parts of operation that you can use as a base for decisions that you need to make in the future.

In terms of your CMMS selection, you may need to know if this form of graphic is available, or if it can be easily produced. There are substantial advantages from being able to drill down on such a graphic. For example, a double click on the breakdowns part may bring up a list of work orders for that day, for that fleet, or you may choose to have this type of graphic on an equipment basis rather than a fleet basis.

Equipment reliability

Another measure that you will be able to apply from the amount of information that you now have in the system is mean time between failures (MTBF). At the equipment or fleet levels, this is not a deeply analytic measure, merely a guide to how the machine, fleet, or group of equipment is performing in a given period. The MTBF measure is calculated as follows:

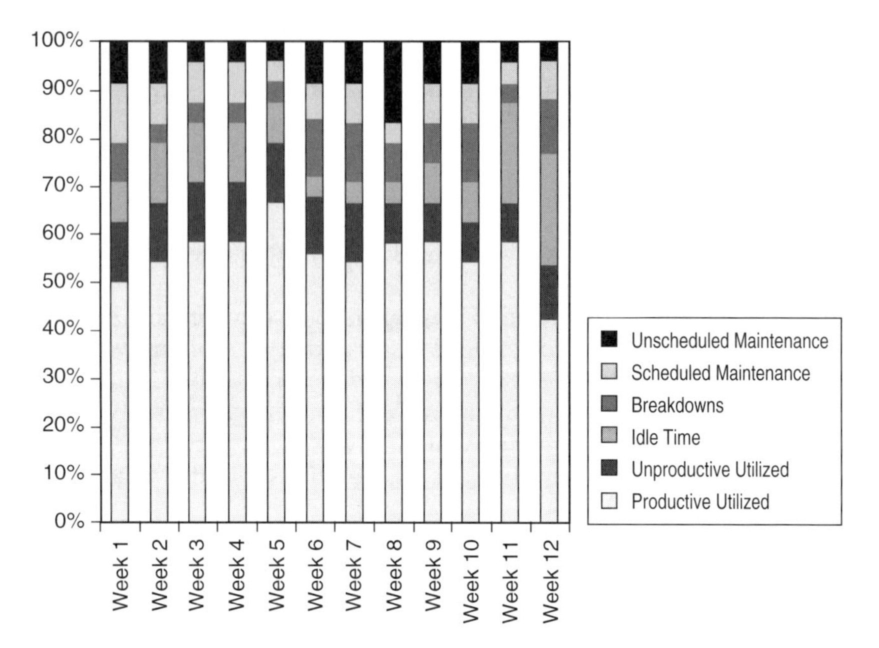

Figure 7.5 Utilization over time.

Productive utilized/number of occurrences of downtime

Here you see the average amount of time you can rely on the equipment prior to having a failure on that equipment. You will notice that the unproductive utilized and idle time codes are not included. This is because, in order to make this measure a fair one and one that you can use as a guide, you need to discount time that you were not using this machine for its productive purposes. Of course, much will depend on how you have defined unproductive utilized as it applies to your operations.

This is a very important measure to determine the real performance of your equipment as seen through the eyes of your primary customers (operations). In addition, it is important to recall that it is easy to have a high availability while having a very low MTBF. Through frequent, short-duration failures, you maintain your equipment at high availability levels while not meeting the expectations or requirements of your operations department.

Equipment maintainability

Total downtime/number of occurrences of downtime

The goal is to gauge the average time for you to recover from a downtime period, at equipment or fleet levels. This also is an important measure because it tells you several things:

- How easy it is to do maintenance to these machines
- How responsive maintenance is to the failure
- Whether you are improving in this function over time

This measure is known as MTTR (mean time to repair) — a very important part of the overall maintenance focus. This measure, however, can be very deceptive to the untrained eye. When a maintenance operation is operating in reactive mode, this number can often be low, indicating that you are very efficient at repairing faults. What it is really telling you in this instance, however, is that you are accustomed to repairing faults; you are good at it because you do it all of the time. As such, this needs to be reviewed in conjunction with the availability and reliability measures mentioned previously.

Overall equipment effectiveness (OEE)

Overall equipment effectiveness is a very harsh but extremely useful measure for evaluating the performance of your equipment. The formula is the following:

$$\text{Availability} \times \text{utilization} \times \text{quality}$$

Built into each one of the sub-components of the formula are the requirements of your plant. For example, availability is one measurement out of a hundred. If you are not at 100%, it will show in the OEE calculation. There are some issues regarding the application of OEE and this should be done at the equipment level as much as possible.

We have not previously spoken of the need for you to evaluate the quality requirements of your plant or equipment. If you are truly interested in evaluating it at the OEE level, then you will need to include this in your requirements statement of the template.

As an example of how harsh OEE can be, consider the following scenario. Imagine your plant with 90% availability. Not a great achievement but perhaps acceptable for your requirements. Imagine you also have a utilization of 90% and an achievement of your quality goals of 90%. Multiplying availability by utilization by quality (90% × 90% × 90%) gives you an OEE of just 72.9%, or 27.1% *less* than achieving the maximum performance of that piece of equipment. The OEE formula is a very good tool for gauging the overall performance of your equipment, but it is not likely to give the results that you want to see at first.

In addition to these reports or metrics, it is good to revise the following information:

- Equipment downtime costs in terms of lost production
- Equipment downtime costs in terms of maintenance costs

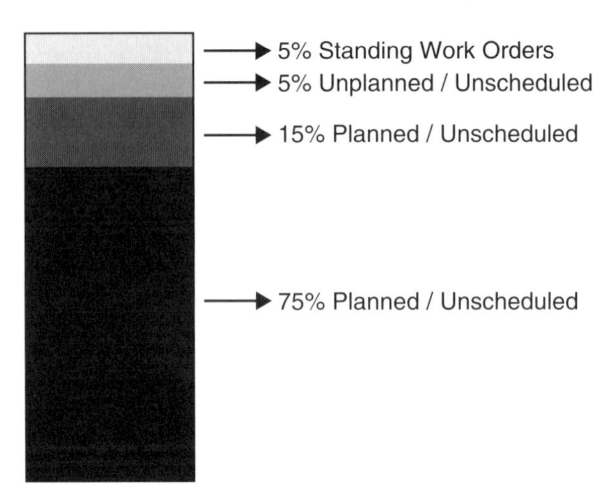

Figure 7.6 World class modes of execution.

With all of these reports, you will have a strong base for the detection of equipment-related failures and enable you to take early corrective actions in this regard.

These are only a few examples of the types of reports available from the CMMS that you need to consider. Using the template approach will guarantee that you have the processes and procedures in place in order for you to deliver these results, and not have to redefine your business processes at a later date in order to accommodate a metric that someone read about in a magazine or on a website. As always, the approach is about being ready and therefore in control of the change, not being driven by it, as is so often the case.

Maintenance process measures

The goal here is to measure how you are doing in the various areas where you operate as a maintenance department. As with all areas of maintenance these measures should be classified into various sub-groups to enable you to locate important issues with greater ease.

Overall measures

Here you need to refer back to the standards of planned/scheduled ratios and maintenance content that were introduced in Chapter Four.

While keeping in mind these two targets for maintenance planning and scheduling contents, the graphics can appear as detailed in Figures 7.6 and 7.7, showing the number of man-hours dedicated either to maintenance content or to the planned/scheduled ratios over time. This allows you to monitor improvements and the effects of any improvement initiatives.

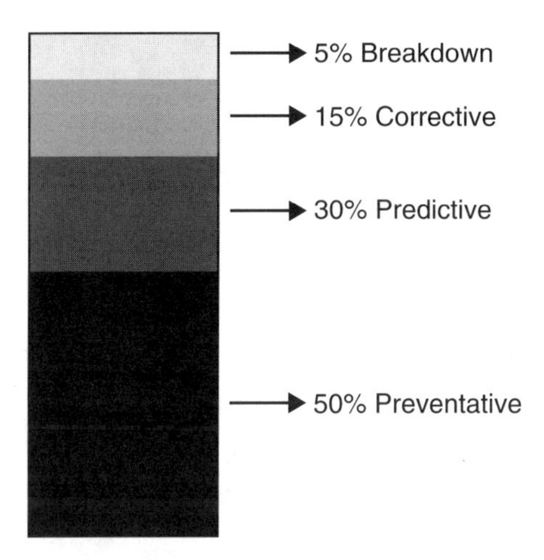

Figure 7.7 World class maintenance program content.

These reports should also be available in a drill-down fashion so as to view the compliance or progress of the maintenance effort by craft as well as the workforce as a whole.

Schedule compliance

You need to be aware of the schedule compliance as well as any reasons for noncompliance at your company. At this point, you can make decisions to either reduce the amount of hours that are scheduled or any other appropriate action.

Other planning indicators

Backlog percentage planned

As stated previously, it is wise to have at least 2 to 3 weeks of planned backlog available in order to have a high level of maintenance preparedness. As such, the formula becomes (depending on the shift arrangements at your plant.):

Total available resource hours × 3 weeks of hours/total resource hours in the planned backlog

Percentage of work orders delayed due to poor planning/scheduling

Here you are trying to understand what the failings of your work order planning system can be. As discussed in Chapter Four, you need to develop a series of work order codes that explain any failings of the work order

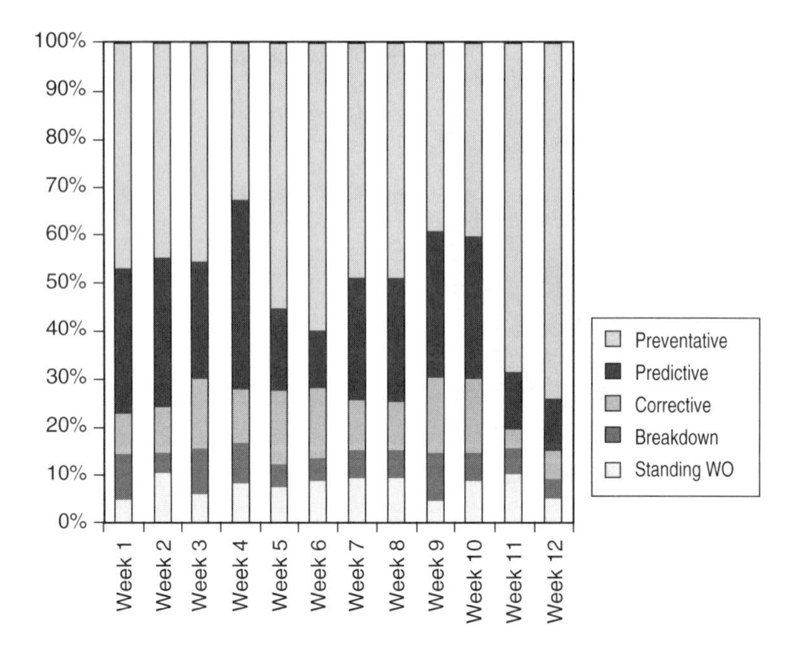

Figure 7.8 Maintenance program content over time.

planning system. For example, excessive delays due to lack of parts may point you to the fact that you are not allowing enough lead time for work order materials.

Similarly, you may be experiencing great delays due to equipment not being available for the work. This would then point you to possible failures in the weekly scheduling process and the follow-up daily scheduling processes.

Work order life by priority

As explained in Chapter Five, you need to set a methodology by which you prioritize work orders, and subsequently plan, schedule, and execute these work orders.

A report showing the number of work orders by age in each respective priority will enable you to give a quick health check to the priorities system. Work orders that are of greater age than they should be for a specific priority need to be actioned quickly while you go about finding out why it is that they were not attended to in the first place.

A result of reports such as these may point to the fact that you are in need of some short-term contract labor in order to get past a critical stage in development, for example.

Although there is a tendency to find high-priority work orders that have outlived their life expectancies, there is also a tendency to find work orders that are of low priority and are not being attended to. The goal is to attend to all of the reasonable and agreed requirements of the operations departments. You need also to take action when faced with this scenario.

Estimations index

As a measure of the accuracy of your task planning, you need to be constantly comparing your results with the real results. Allowance needs to be made here for the fact that you will often experience delays due to one factor or another. However, the total tool time of any work order needs to be close to the estimates that you plan into the work orders. This also needs to apply to materials estimates.

Preventative maintenance (PM)

Percentage of work orders arising from preventative maintenance inspections or services

As a maintenance department, your mission is to gain absolute control over your processes. With measures such as these, you will be able to find out why you are failing in various areas and what can be done about it. The maintenance regime that you use in your plant or facility will determine the performance of many of your assets or pieces of equipment. For example, low ratio of work orders from preventative maintenance routines as opposed to other avenues of corrective maintenance work order raising indicates that your PM regime is not detailed enough.

PM compliance

PM compliance is one of the greater measures that you will be able to apply in this area. As the basis of your reliability growth plans, the PM compliance of maintenance schedules needs to be of a very high level. If it is not, the problem will need to be highlighted and corrected immediately. Poor PM compliance leads to poor machine performance which leads to poor financial custodianship of the company's assets.

Percentage of overtime

As a general guide, you should be able to easily maintain your overtime ratios to less than 5%. Failure to do so may point to a larger workforce or smarter work practices.

Costs charged to standing work orders

As stated several times throughout this book, there are very few occasions that truly call for a standing work order for maintenance work. As such, any costs and materials that are booked to these work orders need to be rigorously monitored to maintain a high level of analytical potential of the work order information.

Reasons for noncompliance

Noncompliance to maintenance schedules can be divided principally into:

- Higher priority work
- Nonavailability of equipment to be worked on
- Nonavailability of resources, both human and material, to do the work

A report showing these three groupings with their sub-classes possibly as a drill down will enable you to easily highlight those areas where you do not have control over your processes, thus wasting maintenance resources.

Failure reporting

Failure reporting must be focused on locating the few critical items that are major stopping points or levers for increasing availabilities and reducing failure frequency rates, e.g., the frequency of a failure code against a critical piece of equipment and the frequency of the same work used to correct a failure. As a part of this, you need to constantly review the number of proactive work orders in the system that are focused on eliminating failures as a part of your continuous improvement focus.

Administrative functions

Percentage of total work covered by a work order

This is imperative to your continued evolution along the path to the predictive state of maintenance. The only acceptable level here is 100%.

Labor and materials as a percentage of total maintenance costs

You will need to see what the content of your maintenance spending is on these items. The goal is that this will ultimately equal 100% of the total maintenance costs.

Contractor usage as a percentage of total maintenance costs

You will need to determine your reliance on contractors and their cost effectiveness. Work orders for contractor labor can and should be contrasted against work orders for the same work using in-house labor in order to detail the differences in costs, time, and material usage.

Cost reports

You should be able to quantify with ease the following maintenance cost statistics.

- *Maintenance unit costs.* Per machine, per fleet, or per process
- *Maintenance budget compliance.* While coming in under budget is good and should be one of your goals on an annual basis, you are nevertheless wasting capital by setting a budget greater than you can adequately use.

Inventory management

Percentage of compliance of items with preset service levels

As stated in Chapter Six, there is a need to set the service levels of specific items depending on their particular characteristics. There is an often-stated view that this should be a flat 95% for all stores items. This is incorrect and is a recipe for bloated warehouses and over stocking on items that may not be needed. However, once the maintenance store policy has been set, according to the requirements for each item, then you must continually review its compliance with those service levels and take action for noncompliance.

Vendor compliance with lead times

You need to review vendor performance, by item if required, on issues such as:

- Late/early arrival of new items
- Late/early return of repaired items
- Compliance with quoted costs

An idea of how your vendors are treating you can give you the ability to either change vendors or reward good vendors with additional work.

Number of urgent requisitions

This should also be reflected in the measures on breakdown maintenance. However, there is often the case where specific departments or people will book out an item to a cost code as opposed to a work order. To account for this, you must review this indicator. Although it is a measure of the stores function, it is very much a measure of the reactive nature of maintenance.

Low usage items reports

For the setting of future stock levels and determining obsolescence in your stores system, a means of reviewing this function is vital. However, you also need to be working with your inventory management policy and revising it in order to determine obsolescence. A part may still retain high criticality even though it is a slow-moving item. The focus of this report is a one year time frame.

Potentially obsolete items

The chance of obsolescence within this range (items with no movement over two to three years) is much greater and as such this report should be included in any inventory analysis regime.

Surplus stock reports

In the hustle and bustle of daily industrial life, there are various reasons why there may be more than maximum limits of a certain store item. However, it is of importance that you focus on these and take action where deemed appropriate.

New items added

The last of your critical inventory reports is the new items added for a period report. Complete with a description and the relevant ABC analysis of the new item, this report can act as a final filter for items that may have slipped through the system. As stated in Chapter Six, the maintenance inventory is one of the greater maintenance expenditures, and, as such, it is fitting that you maintain tight control over it.

Maintenance functional reporting

These are reports that you will need on a daily basis to carry out the functions of maintenance management easily. The majority of them will focus on the backlog system and on being able to pull out variously filtered reports depending on your requirements at that point in time.

As previously stated, maintenance falls in one of three areas: operational maintenance, shutdown maintenance, and technical change management. Therefore, you will need to filter first to determine whether a work order is in one or many of these three categories. I say one or many because a work order may be a technical change that is scheduled to be completed in a shutdown, or it may be an operational work order that has been scheduled for some reason to a shutdown, or any other of the possibilities.

Once you have defined the areas, you then need to apply your requirements of maintenance planning, i.e., you need to determine where an item is in the planning process. For example, you will need to pull up a work order report that contains:

- All items for a specific shutdown that are of a priority 1 classification and are awaiting materials
- All items that are priority 3 classification and require a procedure
- All items for a specific project that are of a safety nature and are planned

The ability to filter and reorganize your maintenance backlog can be either to further the planning effort for a specific area or, once planned, to determine which of the items needs to go into the weekly schedules. It is also a means of revising the backlog content for specific work order classifications and maintenance types.

So, as a general rule, you will need to filter and report on the maintenance backlog in a combination of any of the following criteria as a baseline:

- By area of maintenance (operational, shutdown, technical change, or a combination)
- By equipment
- By originator
- By requestor (if applicable according to the business rules)
- By specific shutdown or merely by shutdown type
- By work order class
  - Safety
  - Maintenance
  - Capital or technical change
  - Environmental
- By maintenance type
  - Preventative
  - Predictive
  - Corrective
  - Workshop repair
  - Breakdown
- By priority: priority-based reporting will determine how you deal with items to schedule and how you deal with planned/unscheduled work.
- By date scheduled: this style of report is very useful for weekly scheduling meetings where you can determine and agree upon the requirements for the next week's work.
- By date raised: this particular report is very useful for the daily, 24-hour work review reports and can be used to determine the priorities for the day and how the daily plan needs to be changed.

In carrying out daily functions as maintenance practitioners, these backlog-style reports will be of invaluable assistance and will aid you in the continuous improvement path that you are on. For example, from the previous listing, you extract a report on the scheduled work, for a specific day, for a specific machine, that has the highest priority. This is one of many useful tools that a planner has for conveying information and for reviewing the work loading for a period or a group of workers.

Parts arrived reports

By being able to see the parts that have arrived for a particular work scheduling period, you make last-minute scheduling calls to determine the validity of your schedule and the need to omit tasks.

Time to go reports

One of the more critical of the equipment monitoring reports that pertain to maintenance planning functions is the time to go report. This report is a list of the major items on a piece of equipment with a calculated value for their change-out date as well as the remaining life in days. From this report, you advise and forewarn suppliers of your major equipment items that this is the anticipated envelope.

In systems and operations that are using condition-based maintenance philosophies and programs, this report may take on a dual function: (1) to indicate the change-out date and the remaining life according to the equipment life statistics normally provided by the manufacturer, and (2) the life expectation figures according to your condition-monitoring readings. You must create this report using the available data on operating hours, the maintenance routines for changeouts of components, and the calculation between the current date, average hours per day, and the remaining life of the component. This is most useful for large equipment items that have long lead times on major components.

Exception reporting

Exception reporting is often overlooked in the scoping of a CMMS as well as in the running of a maintenance department. Exception reporting is exactly that, reporting on all of the exceptions to your business rules and guidelines that you have set for your organization. Following are examples of the sorts of exceptions reports that may be produced, depending on the business processes you have developed and applied.

To maintain high-quality levels of the work in the backlog, you need to filter the work orders created to easily find and react to problems that have been arisen in them. This will assist you greatly in your overall backlog management aims.

Reports need to be set to find work orders with the following noncompliant characteristics.

No priorities

As discussed previously, the priority function is one of the central functions in any system.

Insufficient lines of data

A work order report based on finding work orders with, for example, less than two lines of information. Each must be checked in order to ensure that is sufficient in coverage of the work required.

Noncompliant work order codes

The combinations of work order classes and maintenance types are not a pattern where all maintenance types can be used for all work order classes. As such, you must filter them to find those that do not comply. An example is the class of capital with the maintenance type of breakdown. As capital work is essentially new work, modifications, or purchases, there is no room for this maintenance type.

Priority by age

Although mentioned earlier, this is also an exception style of reporting mainly because it shows up exceptions to your business rules. A work order with a priority 2 rating must be completed within 48 hours. If it has not been done by that time, it is an exception to the rule.

Planned status

Some work orders have been planned, but do not qualify for planned status rating. If you have determined that your definition of planned is that a work order must have resources, materials, procedures, and safety instructions, an exception report focused on the planning status must pick up any of the work orders that do not have these.

Scheduled/unplanned work orders

You cannot allow any scheduled/unplanned work orders to slip through the net at all. You must set a high standard from the beginning. With your horizon of one week in this case, there is no plausible reason that you should have scheduled work that you have not had time to plan. By scheduling unplanned work, you are setting yourself up for inefficient execution of work. You have thrown out the window your drive to eliminate logistical waste and you are beginning to let the rot in.

Corrective work orders closed with no failure codes recorded

This must be highlighted quickly so that you can act on it while the information is still fresh in the minds of those who did the work.

Work orders with minimal or no completion comments

These are vital for your later analysis and other parts of your continuous improvement programs. You also need to capture and act on these quickly so as not to lose the ability to capture the information.

chapter eight

Role definition and training requirements

As a part of the CMMS selection and implementation template, you must also include a focus on the roles that each of the members of the maintenance management team will play in the process. The role of maintenance management practitioners is a very detailed subject that needs to be given an adequate amount of thought prior to the selection and purchase of a CMMS, primarily because you also must be aware of what you will require in terms of training at a later date.

However, as a key part of your requirements specifications, you must consider exactly what it is that you require of the human resources over time, as well as the level of commitment the organization needs to be making to developing long-term relationships with its human assets. This goes to the heart of the corporate goals and aspirations. There are organizations that do not consider a long-term relationship with employees a valuable commodity. These organizations often resort to contract labor solutions while enforcing their procedures and processes. Others feel that it is a great benefit to have long-term resources with a long history with the organization. They are committed to creating workers of value that have definite career paths; these organizations have very definite measures of employee performance, in line with corporate goals.

Neither approach is the "best" way to proceed. Each organization must make decisions in this regard, depending on the corporate needs and goals. A contract labor force may be the only alternative for a plant that, for operational reasons, is located in inhospitable surroundings or very far from the luxuries of modern-day life. In some areas or industries where high turnover is a part of life, there is still the feeling that training of workers only provides other companies with trained resources. In any case, there is a need to manage this within the CMMS and as such it forms a part of your requirements template.

The human resources component

During this phase, you must focus on the role of human resources within your operations and how you can lever this function to assist the maintenance efforts. As a part of the corporate development of your organization, the human assets should be part of your primary areas of attention. By attending to the requirements at the this level, you can maximize your use of the employees that you have. Where does this all begin? Before considering anything else, you must consider and possibly develop the profiles of each of the positions within your maintenance organization.

Beginning at the craft level, what are the main craft distinctions that you will require to operate proficiently? Once this is defined, you then must look at what the base requirements are for a craft employee entering your company. From there, what are the profiles required for a maintenance planner, supervisor, and superintendent manager? Can your craft employees be developed to ultimately qualify for positions such as these?

Once the new employee has joined your operations, what do you do with him? What are your end goals with this employee, and how will you measure his progress toward these end goals?

From these principal questions, you can establish the needs for various key functions of your integrated human resources system. First, you must establish the mix of skills and experience that qualifies a person to apply for each of the positions within the organization. To be able to manage this, you will need some manner of recruitment functionality and a way of following applicants through that process.

You must be able to set career goals for them once employed. The ultimate goal is to develop a resource that is utilized in various situations and able to perform with efficiency and the utmost attention to safety and quality, giving you the ability to improve the overall effectiveness of your workforce bit by bit. There is a need to set a career plan, possibly linked to pay scales. One of the elements that you must focus on here is the possibility of cross-training maintenance employees to perform basic to medium-level complexity tasks in other craft areas.

There are obviously a great many industrial relations issues to be considered here as well as safety and workforce-level issues. Any move to do so must be handled not only with the utmost discretion but also with a very high level of detail given to ensuring that any cross-training is of a high level of quality and is beneficial to the organization.

As a part of the management of career paths, there also must be the management of training periods within the system. Without inclusion of some form or another of training time an employee will be required to undergo, your entire capacity scheduling process is in jeopardy.

Once the career path has been developed for an employee, you then must consider how the employee will progress down this path and how you will factor in future wage or salary calculations. As such, some form of evaluation process must be recorded with the effects of this being translated

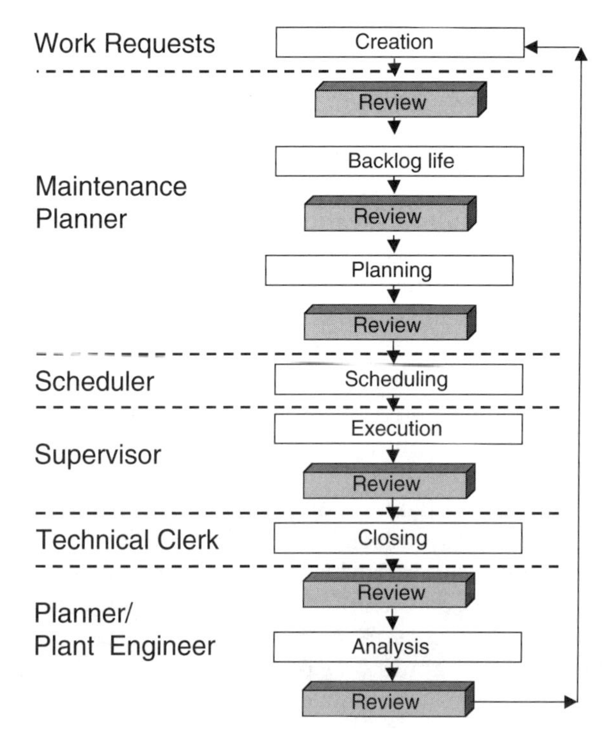

Figure 8.1 Work order life cycle with roles.

into both future training requirements and future pay increases. This area is a difficult one and there are various methodologies and systems on the market that are focusing on it. Whatever system or approach is being undertaken, it is important that the CMMS will support this function.

With this basic human resources function taken into account, you are able to lay the groundwork for a maintenance organization that will be both well developed and enduring. Although often overlooked, the importance of the this cannot be undervalued.

A final element that is at times grouped into the human resources functions is setting the corporate hierarchy. This must be a possibility in order to be able to manage and deal with the various authorization processes and limits that can exist. If a supervisor needs something, he should be able to order it; if it is outside of his authorization limits, then the purchase order should pass immediately up the corporate hierarchy for authorization.

The ability to do this will put safeguards into the materials ordering and purchase order creation side of the business. It must be structured with a view to maintaining control while allowing flexibility. There is the tendency among authoritarian-style corporate structures to create bottlenecks in the fluidity of workflow by requiring authorization for all elements ordered.

Another aspect that may be of use it that of creating more than one corporate hierarchy and reporting structure. This, as will be explained later in this chapter, is very useful for the application of a different structure during shutdown periods, for example.

The maintenance profiles or roles

Earlier in this book, we looked very briefly at the work order life cycle. At that point, it was to determine the various parts of the maintenance processes that are important to your development toward the planned state of maintenance. Now you will be reviewing it again with the view of detailing further the roles required to do specific parts of it.

As mentioned before, this is a typical work order life cycle, and it may vary according to each specific requirement. However, the basic principles remain the same. The work order life cycle will be applied to all maintenance areas whether operational, shutdown, or motivated by technical change. All work order life cycles must contain the periodic reviews and revisions of information to maintain a high level of quality in the backlog, and ultimately a high level of quality to the execution stages.

The first part of your work order life cycle is that of the work requests. In this example, you will use all crafts and all operations personnel to raise work orders, using the concepts of work filtering via the work request system, and leaving the maintenance planning and management to the parties responsible for the creation, management, execution, and analysis of the work orders.

The second phase in the work order life cycle is that belonging to the maintenance planner. There is still a disproportionately high number of installations that do not use maintenance planners, or when they do use them, they are not utilized in the manner that best befits the role. Many maintenance planners can end up as clerks or "gofers" for supervisors. This is not the role that the position is designed for. The position is the section's "know it all." Maintenance planners must have all of the equipment details plus history and general statistics at their fingertips at any given moment. They are the enforcers of the maintenance systems and, given the correct amount of support, the lubricant that assists a maintenance department to run smoothly and achieve high goals. They are also critical elements in the setting and revising of maintenance strategies, and must form a part of any continuous improvement asset management planning.

As can be seen in the work order life cycle, the role of the maintenance planner in this situation is to perform the functions required of them that pertain to backlog management, planning, and reviews to ensure the consistent, high level of information in the system.

Outside of the work order management system, there are other functions that the maintenance planner undertakes or oversees as part of their role as the custodians of the system's information reliability. This may include control of the tracing of high-cost or critical equipment, and management of the repairables management program in conjunction with the stores department.

The third part of the maintenance work order life cycle covers the role of schedulers in the maintenance process. Some plants do actively use schedulers as a separate role and have achieved good results. Others do not use them and also have achieved good results. The bottom-line requirement is the size of the installation and the number of people that the maintenance department is required to actively manage at any given time.

The scheduling function of the work order life cycle is to deal with the logistical aspect of work execution, i.e., the equipment, materials, and human resources available at the time that the work is to be executed. Although not stated here, this function also needs to take responsibility for follow-up during the week to ensure that daily changes to the schedule are done in a controlled and adequate manner in order to maintain control over the maintenance resources and systems of execution. It also falls to the scheduler to prepare and distribute the work packages as mentioned in a previous chapter.

The next two items are associated with the execution and then the closing of the works order. These are assigned here to the supervisor and the technical clerk, or maintenance clerk, respectively. The supervisor obviously needs to fulfill his role in the process by attending to the items associated with the execution of the work. The maintenance clerk needs to be the primary interface with the system after the required quality checks have been done to ensure that the work orders contain the important and sufficient data.

The last area of the work order life cycle is the analysis of work done. In the section on reporting, we looked at the types of reports that were available for this type of analysis. The goal here is the improvement of equipment performance and of maintenance execution performance. In the life cycle graphic (Figure 8.1), the roles assigned to this are the maintenance planner or the plant engineer or both. This is only a suggestion. All analysis should be done by anyone in the organization, from financial controllers to supervisors, in order to best highlight and decide on solutions to the two areas mentioned. The analysis of equipment performance also needs to involve revisiting the RCM or equipment strategy base to ensure that the PM regime is adequate for the performance required. It is also advisable to incorporate some form of RCFA technique as a means of possibly eliminating the need for maintenance.

Training matrix for CMMS implementations

With the basic roles and participations of the various members of the maintenance team now determined, you are able to move to the creation of the training matrix that you will implement during and after the CMMS implementation period. One of the key elements here is the ability to have one or two super users per group. The end result is that you should be able to continue training people after the implementation is completed and the consultants have left. It is important to maintain a degree of independence in order to reduce future dependence on outside resources.

The training matrix is designed to determine the training that is required for each role in order to best accommodate the functions that we have been describing thus far. And although much of this will center on learning the new CMMS, a great deal of this should also be focused on learning the new or redefined business processes and structures. This should include one-on-one meetings between department heads and their staff in order to discuss the role descriptions and what is expected of the employee in the future.

You end up with at least two training matrices to highlight the training that is needed — one for the system that is to be implemented and one for the processes that are to be implemented along with it.

It is also important to realize that this is a part of the implementation template at the stage of definition. At this point, it is a guide to what you think will be required. When the product to be implemented has finally been chosen, this may change somewhat in consultation with the software vendor.

However, in a multi-site implementation, the matrix will prove critical to be able to ease through later implementations. It will also provide a guide for the training requirements for new employees in the maintenance department.

The maintenance system training requirements

Here you are able to see the system training requirements for each member of the maintenance team, according to their specific roles. You will also notice that at the bottom of this table, there is a box marked "Outside of the Maintenance Department." This refers to inventory controllers, human resources, and other roles that may have the requirement to interact with maintenance. One of the prime areas is in the raising of work requests.

The matrix may be more detailed in order to train people only in the area required to perform their role and not the entire part of the CMMS. For example, the requirements to understand work orders will differ between the maintenance manager and the maintenance planner. The planner will use this system to perform extensive reviews and advanced backlog management, whereas the maintenance manager may only raise work orders and review them in a high-level and general sense.

Depending on the depth of the CMMS that you purchase, this may be necessary to apply to all of the areas that are being implemented. For example, advanced inventory management functionality will be detailed as to who does what as part of the processes, and, most importantly, the human resources portion of the system will be extensively explained and transmitted to all managers. Part of their interaction with the system will be updating worker performance appraisals, booking people on course, and even editing career planning information. As such, the ability to cross-train the various managerial streams is of vital importance here.

	Work Requests	Work Orders	Asset Register	Rebuildable Items	Store Requisitions and Purchase Orders	Maintenance Scheduling	Predictive Maintenance Routines	Time Entry
Maintenance Manager		X	X		X			
Maintenance Planner		X	X	X	X	X	X	
Maintenance Scheduler		X	X	X	X	X	X	
Maintenance Supervisor		X	X	X	X			X
Operations Supervisor	X		X		X			
Craft Employees	X				X			X
Maintenance Clerk	X							X
Plant Engineer	X	X	X	X	X	X	X	
Outside the Maintenance Department	X			X	X			X

Figure 8.2 CMMS example training grid.

The maintenance process training matrix

In the maintenance process training matrix, you will notice that the emphasis is different. It focuses on the use of the modules for the processes that you are developing. If the work order prioritization system is going to function well, there must be a full level of understanding among all personnel raising work requests and work orders. It is not anticipated that this area of the training requirements will contain a long course; training will probably require periods of no more than one day. However, training periods are vital elements of a CMMS implementation that is to be fully successful.

Again if the scope and functional depth of the CMMS provides for advanced management of other areas of the corporate environment, this may must be repeated for each of the areas undergoing the implementation procedures. The Human Resources department will define how the corporation sets career plans, and will determine the acceptable level of training for each of the roles in the company. The supply and warehouse departments may have specific requirements for their processes when dealing with vendors of repair services. This will also need to be transferred.

Along with the training matrices involved with the overall implementation and use of the CMMS, this may be a good time to evaluate the training that is required for the maintenance practitioners within the organization. Here you may be able to highlight maintenance planners specifically to give an example.

In an environment where many organizations still do not use maintenance planners, and in those that underutilize them, how will you know what it is that a planner does? And further to the point, how is the poor old planner to know what their role is in the process?

	Work Initiating	Back Log Management	Prioritizing	Work Order Management	Work Order Coding	Planning and Estimating	Capacity Scheduling	Analysis Techniques
Maintenance Manager			x		x			x
Maintenance Planner	x	x	x	x	x	x	x	x
Maintenance Scheduler	x	x	x	x	x	x	x	x
Maintenance Supervisor	x		x	x	x	x	x	x
Operations Supervisor	x		x	x	x	x		x
Craft Employees	x		x	x	x	x		
Maintenance Clerk	x		x	x	x			
Plant Engineer	x	x	x	x	x	x	x	x
Outside the Maintenance Department	x		x		x			

Figure 8.3 Maintenance process training grid.

There are a great many areas where you are able to gain training for maintenance planners and other practitioners. Courses such as Front Line Supervision and Estimating and Job Slotting are only two of the possible list of examples. These types of training courses can be done via correspondence or distance education as well as through a multitude of consultancies and CMMS vendors.

Defining the role descriptions

One of the more common causes of employee frustration and political infighting is the lack of clearly defined boundaries between roles. Part of the focus of the work order life cycle was to bring you to this point. But you also must be considering what you are going to do in this situation, and how you are going to go about it. In addition, you must define where exactly in the corporate structure you need roles defined.

The role and reporting functions of a maintenance planner are distinct from the role and reporting functions of a shutdown planner, but these roles and functions may be handled by the same person. In this case, how do you go about this? The answer in the case of an organization that uses its internal resources to manage the maintenance shutdown is to have two distinct role descriptions for the same person. One description determines the everyday responsibilities and duties of the person in his normal role, while the other determines what he does under extraordinary circumstances such as a shutdown process. Although we will not discuss it here, one of these extraordinary circumstances may well be crisis or disaster management.

Other roles also must be detailed in such a manner. During a shutdown, due to the large number of activities and of people that may be involved, there may be a need for a permit or authorizations controller. This role may not exist in the day-to-day operation of the company. As such, the role description needs to be written and the role needs to be assigned as a specific employee's responsibility during the shutdown period. However, not all roles will be handled this way. The role of a supervisor, which is focused on front-line leadership, may not change dramatically during the shutdown. The role may be written in a manner that includes the shutdown responsibilities because they do not change greatly.

As a general rule, the contents of a role description must be detailed and focused on all of the responsibilities of that role, as determined by the process definitions that you have now written. The main content areas of a role can be described in the following manner:

- *Main role focus:* Why the role exists and what the overall general concepts are.
- *Safety responsibilities:* Due to the high degree of importance of this area, it is important to highlight the responsibilities specific to each role.
- *People responsibilities:* As we all know, people skills can make or break an organization. Skills such as communication, ability to work as a team, etc., are needed to advance the maintenance organization. As such, it is necessary to list these as distinct from the main role focus and safety responsibilities.
- *Role descriptions per business process:* Here you place some attention to the business processes this role will be involved in and its specific role in that process. This needs to include time limits and other details.
- *Role interrelationships:* As stated earlier, there is a lot of harm done organizationally by lack of definition of the parameters of each role. I personally have experienced situations where I may have had three or more superiors each giving me conflicting instructions. Here the intention is to identify the other roles that interact with this role, as well as how they interact and what the corporate relationship is.
- *Role deliverables:* In addition to the responsibilities and accountabilities listed previously, it is also a good thing if you are able to list the main deliverables of this role, i.e., when done successfully, what do you expect from the role? As a part of a corporate culture that uses career planning and employee evaluations, these measures can help to separate the emotional or human judgement aspects during an evaluation. The deliverables of a role may not apply only to that role but to various roles. For example, the lack of rework after a maintenance shutdown is a shared deliverable of many roles involved in the process.

There are two example role descriptions contained here. One for a maintenance planner during an operational maintenance and one for the same person during the shutdown period. Although apart from the implications to the hierarchy creation and training and authorities matrices, there is no direct link to CMMS implementation. However, as mentioned in the first chapter, this is a period of opportunity to redefine the business in order to optimize its function. Role descriptions are a fundamental part of this. The goal of including these is to provide a baseline for personnel currently developing roles descriptions. The roles refer to different role descriptions; some you may or may not have mentioned. In order to make this align more with the majority of organizational requirements, it is assumed that the role of the maintenance planner and that of the maintenance scheduler are combined.

Shutdown planner

Main focus of role

The shutdown planner is a key figure in the planning, quality control, and fluid execution of the shutdown. As the guardians of specific knowledge related to tasks to be completed, shutdown planners act as advisers to supervisors, team leaders, and craft employees. This is central to shutdown success.

Responsibilities

Safety

Actively promote a culture of creation of safe work procedures and of safety awareness.

Support initiatives of the safety officer and ensure team awareness of these.

Ensure that work orders have additional safety information written into the long text to better provide information to those required in the executing of a task.

People

Actively promote a focus on shutdown systems as well as a culture of quality workmanship.

Provide input to assistant coordinator as well as technical input relating to additional scope that they want included.

Pre-shutdown role

Effectively manage the backlog for their areas of control to ensure:

- All work orders are resourced accurately.
- All jobs meet the criteria for shutdown inclusion.
- All work orders included in a shutdown are coded with the relevant revision number.
- All work orders have the duration of the task entered in an accurate manner.
- As much as possible, all work orders, including PM tasks, are to be highlighted and coded prior to the closure of the scope for a shutdown.
- Any additional works raised are to be assessed against the guidelines of "nice to have" or "essential."
- All work orders are to contain additional instruction in the long text to ensure that tasks are completed to a level of quality consistent with longevity of the repairs or replacement and will negate the need for rework.
- Effectively communicate between the planning team members to ensure that the "big picture" is understood by all.
- Ensure all required parts will be on-site to tackle the tasks during the shutdown. Highlight long lead items and act on them early to ensure delivery. Liaise with stores and purchasing officers to do this.
- Develop the logic as well as review the resources and durations of each task within their area of control.
- Relay information to assistant coordinator after plan handover to ensure that scope is accurately defined.
- Assist in highlighting and developing the required safe work instructions or task instructions for carrying out tasks as required.
- Promote a culture of adherence to shutdown systems and procedures. Assist in raising understanding of why certain items are critical, such as accurate labor hours accounting, strong job card history, etc.
- Ensure adequate tooling for tasks is available via liaison with supervisors and front-line employees. Feed this information via the assistant coordinator for their action.

Shutdown execution responsibilities

- Attend daily status meetings.
- Conduct walk-around inspections to determine that all work for areas under their control is:
  - Adequately resourced.
  - Adequately supplied with parts required.
  - Being carried out in the manner required.
  - Being carried out to a satisfactory level of quality.
- Advise assistant coordinator and shutdown coordinator of any additional tasks uncovered during execution and the implications of doing (or not doing) them. Consideration is to be given to:
  - Safety implications.
  - Manning and cost implications of inclusion.

  - Cost implications of exclusion.
  - Duration implications.
- Raising and coding of all work orders representing tasks uncovered during a shutdown. All work orders created during the shutdown are to be coded with the revision number and with the maintenance code CA (corrective actions).
- Actively promote the culture of control to be established by performing snap audits of job cards and discussing the value of accurate data with front-line employees.
- Actively look for areas of improvement where processes or tasks can be made more efficient.

Post-shutdown responsibilities

- Assist in the closeout of work orders to ensure there is accuracy of data in the system.
- Assist in the collection of outstanding job cards to ensure that all data is recorded.
- Supply post-shutdown report highlight issues during the shutdown. For example:
  - Things learned.
  - Things that can be done better.
  - Things that were done well.
- Participate in review meeting and carry out actions tasked from this meeting.
- Ensure that all unused items are credited back to the store system to reflect accurate stock levels.

Relationships

The shutdown planner is a pivotal position allowing a detailed focus on a smaller area of a plant to be undertaken. The key deliverable of this role is in the work order management and quality assurance function related to their specialist area of the plant. The role interactions associated with this position are:

Supervisors/team leaders – co-worker. Provide as much information as possible to ensure safe and quality-driven execution of tasks. Work with the supervisor during shutdown periods to ensure all data collection functions are performed to the minimum level required for analysis.

Assistant coordinator – co-worker. Provide accurate details regarding scope, scope creep, durations verification, labor resourcing, and logic setting.

Shutdown coordinator – superior. Provide support as directed during shutdown periods. Coordinator is to be supplied with accurate information regarding the execution of the shutdown at all times.

Safety officer – co-worker. Support the creation, implementation, and promotion of all safety systems proposed and managed by the Safety Officer.

Shutdown manger – superior. Follow directives as applied.

Permit controllers – co-workers. Provide information related to permitting or preparation of work areas, initially via long text entries on job cards and ultimately in detailed discussions prior to plant shutdown.

Measurables

The role of the shutdown planner is to be measured against:

- Performance of all tasks in a safe and timely manner.
- Active support of safety, system, and quality initiatives.
- All tasks for their area of control to have adequate labor resources assigned.
- All tasks under their area of control to have adequate parts to do the task, barring uncontrollable situations.
- Rapid identification of additional tasks and provision of risk analysis of them for inclusion.
- Provision of post-shut report
- Lack of rework immediately after a shutdown, showing good-quality instruction and follow-up discussion with supervisors/team leaders.

Operational maintenance planner

Main focus of role

The operational maintenance planners are key figures in the planning, quality control, and fluid execution of all operational maintenance works. As the guardians of specific knowledge related to tasks to be completed, their role as advisers to supervisors, team leaders, and craft people executing a task is central to success. They provide support to the workforce via adequate planning, scheduling, and data control of work order information, and by the revision and updating of maintenance strategy information and asset management information.

Responsibilities

Safety

Actively promote a culture of creation of safe work procedures and of safety awareness. Support initiative from the safety officer and ensure team awareness of these.

Ensure that work orders have additional safety information written into the long text to better provide information to those required in the executing of a task.

People

Actively promote a focus on work requesting, planning, and data execution systems as well as a culture of quality workmanship.

Provide support information to supervisors, managers, plant engineers, and craft employees, as required.

Backlog management and planning role

Effectively manage the backlog for their areas of control to:

- Ensure all work requests meet the requirements of the corporate work order criteria.
- Ensure work requests are actioned within a reasonable time frame (as determined by priorities).
- Ensure all work orders are resourced accurately.
- Ensure all work order priorities are correctly assigned.
- Ensure all work orders to be included in a shutdown are coded with the relevant revision number.
- Ensure all work orders have the duration of the task entered in an accurate manner.
- Ensure all work orders have the material requirements defined and ordered.
- Ensure all work orders are assigned to the correct work team, contract group, or person.
- Promote a culture of adherence to planning systems and procedures. Assist in raising understanding of why certain items are critical, such as accurate labor hours accounting, strong job card history, etc.
- Ensure adequate tooling for tasks is available via liaison with supervisors and front-line employees.

Scheduling responsibilities

- Attend daily status meetings.
  - Provide and review 24 hour reports.
  - Adjust schedule to accommodate higher priority tasks.
- Manage weekly scheduling meetings.
- Prepare capacity resource schedule.
- Ensure all logistics are taken care of:
  - Equipment availability.
  - Resource availability.
  - Material arrival dates and times.
- Prepare job packages for execution.
- Actively look for areas of improvement where processes or tasks can be made more efficient.

Execution responsibilities

- Provide additional information when required to assist in the fluid running of the execution process.
- Actively promote the culture of control to be established by performing snap audits of job cards and discussing the value of accurate data with front-line employees.
- Receive completed work orders from the supervisor, revise, and pass to the maintenance clerk.
  - Review for accuracy and quality of information.
  - Additional tasks requiring work orders and planning.
  - Changes to work order templates.

Asset management responsibilities

- Update/create maintenance strategies as required.
- Update/create work order templates as required.
- Ensure data for equipment tracing is entered to the system.
- Ensure items are either sent off-site or repaired on-site within adequate time frames.

Relationships

Operational maintenance planner is a pivotal position providing the planning, scheduling, and control of work order quality to ensure efficient work execution.

Supervisor – co-worker. Provide as much information as possible to ensure safe and quality-driven execution of tasks. Work with the supervisor to ensure all data collection functions are performed to the minimum level required for analysis.

Safety officer – co-worker. Support the creation implementation and promotion of all safety systems proposed and managed by the safety officer.

Maintenance manager – superior. Follow directives as applied.

Measurables

The role of the operational maintenance planner is to be measured against:

- Performance of all tasks in a safe and timely manner.
- Active support of safety, system, and quality initiatives.
- All tasks for their area of control to have adequate labor resources assigned.
- All tasks under their area of control to have adequate parts to do the task, barring uncontrollable situations.
- Lack of rework.

- Efficient management of repairable parts.
- Accurate tracing of components.
- Participation in corporate continuous improvement initiatives.

chapter nine

Compiling requirements

The implementation template

After detailed consideration of all of the elements of the CMMS and how they can be managed at your company, you have arrived at the area where the CMMS template is assembled. The template is the result of all the work you have done to date, and forms the request for quotation (RFQ) document that you will have to submit as part of the ongoing selection process.

One of the vital points is to be prepared to be disappointed. There are only a few top-range systems that contain all of the functionality that you will require, and you will need to search for them. As mentioned in the first chapter, there may be a need to look at the creation of a best-of-breed system for your own specific requirements. Through the advances in integration and open software architecture, this is becoming easier to do.

However, one of the important points to bear in mind is the following: Although you are buying software, and the resulting project will focus on its implementation and use, never lose sight of the fact that this is a maintenance project, not a software project. Do not allow it to get hijacked by the software department.

At all times throughout the template implementation, make reference to documents that are created to manage the process that is being spoken about and attach them to the template document. Quite often when receiving RFQs or requests for tender from companies, there is a great deal of writing regarding what their current processes are but no specifics as to what is required.

1. Maintenance areas
   a. Must be able to manage operational maintenance in the form of regular preventative and repetitively programmed tasks as well as general corrective tasks
   b. Must have the ability to filter tasks associated with technical change management and manage these as detailed in the attached document

   c. Must have advanced shutdown management capabilities and the ability to manage shutdowns in the manner stated in the attached document; also requires definition of shutdown types
2. Maintenance processes
   a. Must contain a work requesting function capable of telling who, what, when, and why the work is required
   b. Must contain a form of defining the work order class as well as the maintenance type being performed
   c. Must have a way of identifying priorities as defined in the attached document and also be able to filter them
   d. Must have a planning status indicator function and possibly advanced forms of auto-managing this function
   e. Must contain an easy, online capability for managing maintenance backlog and for filtering this so that the maintenance backlog can be seen per user requirements
      i. By area of maintenance (operational, shutdown, technical change, or a combination of all three)
      ii. By equipment
      iii. By originator
      iv. By requestor (if applicable, according to the business rules)
      v. By specific shutdown or merely by shutdown type
      vi. By work order class
         1. Safety
         2. Maintenance
         3. Capital or technical change
         4. Environmental
      vii. By maintenance type
         1. Preventative
         2. Predictive
         3. Corrective
         4. Workshop repair
         5. Breakdown
      viii. By priority
         1. Priority-based reporting will determine how you deal with items to schedule and how you deal with planned/unscheduled work
      ix. By date scheduled
         1. This style of report will be very useful for weekly scheduling meetings where you can determine and agree on the requirements for the next week's work
      x. By date raised
         1. This particular report is very useful for the daily 24-hour work review reports and can be used to determine the priorities for the day and how the daily plan needs to be changed

f. Must contain the ability to schedule PM tasks
    i. By day or calendar period
    ii. By statistical function
    iii. By a combination of statistics
g. Must contain the ability to manage corrective tasks in an easy manner
h. Must be able to manage predictive routines and resultant work
i. Must be able to manage scheduled component replacements
j. Must contain work order template functions
k. Must contain maintenance material lists
l. Must contain the ability to perform capacity scheduling in accordance with the attached document
m. Must contain the ability to easily reprogram tasks
n. Must contain the ability to report on skipped maintenance routines
o. Must contain the ability to program graphically or to interface with a Gantt chart application
p. Must contain the ability to use smart failure codes to analyze equipment performance
q. Must have the ability to capture completion comments from each work order completed
r. Must have an advanced equipment register capable of creating a hierarchy of equipment and containing all of the details that your organization requires
s. Must have the ability to monitor and create work orders on alarm levels according to condition-monitoring requirements
t. Must be able to easily program complex service periods and routines
u. Must have the ability to contain meter readings from equipment
v. Must be able to change routine frequencies or types with changes to the PF (potential failure) interval of tasks

3. Inventory management
    a. Must contain functionality to handle a detailed, searchable stores catalog, according to the attached document
    b. Must be able to create requisitions and purchase orders
    c. Must be able to produce simulations for analysis
    d. Must be able to interface or integrate with online marketplaces and vendors' sites
    e. Must be able to manage and classify items according to the ABC methodology in the attached document
    f. Must be able to manage space within the warehouse
    g. Must be able to manage multiple warehouses
    h. Must be able to manage repairable items in the manner attached
    i. Must be able to manage traceable items in the manner attached

4. Human resources
    a. Must be able to manage personnel details
    b. Must be able to manage training requirements
    c. Must be able to manage career planning
    d. Must be able to manage advanced shift patterns
    e. Must be able to manage worker performance appraisals and link this to possible training suggestions
    f. Must be able to manage planned and unplanned absences from work
    g. Must be able to manage complex employee classifications and sub-classifications
    h. Must be able to contain multiple hierarchies with corresponding reporting and authorizing structures
5. Financial reporting
    a. Must be able to produce reports on the financial performance of equipment by
        i. Accounting period
        ii. Unit costs
        iii. By process and by equipment
6. Reporting
    a. Must be a flexible reporting system able to modify and change per the organization's requirements
    b. Must support the availability structure in the attached documents, therefore supporting utilization, quality, and other measures that are required
    c. Must be able to produce complex backlog reports as per the attached document
    d. Must be able to produce planned/scheduled ratio reports
    e. Must be able to produce "time to go" reports for equipment component life expectancies
    f. Must be able to produce OEE reports
    g. Must be able to produce the exception reports listed in the attached document
    h. Must be able to produce the PM compliance report
    i. Must be able to produce the priorities by age report

With definitions such as these, along with the attached documents that are mentioned, there will be no doubt as to what your organization requires. Having gone through the exercise of redefining the business processes that your organization will use, you are very much in control of the process, and driving the system.

There are, however, additional considerations that need to be weighed at this time which revolve around the capabilities of possible software providers and other system-type issues.

Other considerations

Although you now have the majority of the functionality requirements detailed according to your business requirements, there are yet other factors that you should be considering.

Price range

Having determined the possible savings and return on investment, you now know the price range of the software that you are looking for.

Hardware requirements

This is often a key issue: What are the hardware requirements of running a system of the size you are considering?
Will it run on a standard desktop PC?
Do you require a server? If so, what size?
What are the communications issues you face? Is there a need for fiberoptic cabling?

Software platform

What is the software platform?
Is it a difficult area to find skills if you need to make changes?
Will there be training on this area of the system?
Are there other tools you can use for ease of interfacing and uploading vast volumes of data?
Is it Internet enabled? Internet based?
What are the configuration requirements? Are they difficult or extremely complex?

The company

How old is the company?
Does it have a good track record regarding on-time and on-cost delivery?
Is there a feeling of overpaying for the name alone?
Can the company provide evidence of ROIs achieved elsewhere?
What do the company's clients say? Are there references?
What are the skill levels of the functional and technical consultants?
Is the system well known? For example, will there be issues finding people to manage the system after a few years?
Is the company financially stable?

Transfer of information

Is the software company contemplating the creation of migration programs to the new system?
Are they capable of creating the interfaces that you will require in order to have a working system with the ability to monitor all systems and processes?

This is a very thorough system for the creation of a selection template with which the vendors will need to align themselves. From this point, the commercial process of selection needs to be thoroughly controlled so as not to overstep the time limits for this part of the project.

chapter ten

The project

We are now at the final part of the CMMS implementation — the implementation itself. This is the most demanding part and will test the limits of the resources dedicated to it. In Chapter Two, we set out the team that will oversee the implementation from its inception to its post-implementation stage. The team is now the driving force during the implementation phase.

At all times during this phase, be sure that you seek the guidance of the company that has provided the software. They are experts at this part of the process. However, maintain the leadership position and demand compliance with the business process statements that were set out in the selection document and that the software company agreed to perform.

From the beginning of the project, it is always important to take into account the common failures of CMMSs and to direct your attention to making sure that these do not occur in your implementation.

Failure 1: Lack of corporate support

This is the most critical of all of the failure areas. Support, direction, and assistance from upper management is the one thing that will ensure the success or failure of any CMMS implementation. Often there are cases where upper management will delegate authority to one or two of the project leaders, such as the project director, and it will then wash its hands of any involvement. These cases usually end up with unhappy clients and unworkable solutions.

The corporate leadership needs to be involved, and perceived as involved, at every step of the project. They need to make sure that the organization is aware of what is happening in the project, and that they are expected to be involved in this great investment that the company is making. This can be done a myriad of ways, although the most useful is always personal appearances in support of the system and the project.

Failure 2: Lack of adequate license purchases

A sure-fire recipe for a disastrous project is not purchasing adequate licenses. At the end of the day, there will be such a small impact on the corporation as a whole that it would have been better had the investment not been made.

Failure 3: Lack of adequate training

Although we have set out the system and process training in the chapter on training, this is not the end of the matter. There also remains the fact that training will be passed to a great many, if not all of the company's employees either in the form of training on specific parts of the system, on general processes, or merely as a measure of awareness.

One of the common errors in this respect is allowing the staff to conduct the training without sufficient preparation, time, or experience. A train-the-trainer-style course is an essential element of any moves in this manner. There have been many failures in the CMMS world due to the trainers not knowing how to train end users.

Failure 4: Lack of history in the system

There must be a method of migrating all of the data from your current system. I have seen major organizations that changed systems many times in a short period with the end result that all of their maintenance history and maintenance routines were lost. Suddenly a plant with 20 years of operations had no maintenance routines at all!

Failure 5: Lack of adequate use

This goes to the heart of failure number 1. Without adequate use of the system, there will be no advances after its implementation. All manual, workaround-style systems should be banned and actively discouraged throughout the organization. There can be no room for allowing political infighting to wreck this project, especially after large sums have been spent on it.

Failure 6: Lack of adequate test procedures

Prior to the go-live of the system, all of its functions must be tested in a thorough and intensive manner by the end users. This cannot be skipped.

During testing, several areas will require attention:

- The functional operation of the system itself (does it do all that it should?)
- The function of all interfaces that have been purchased
- The functionality of the migration
- The functionality of additional programs that have been purchased

Setting out the project timeline

As with all well-run projects, this project should be pre-scheduled and tracked rigorously to ensure that all valid milestones are being hit and that there is as little as possible scope creep and changes during the implementation phase. A typical timeline may include the following segments as a guide.

- *Project planning phase:* Corporate leadership workshops on the focus of the project and what the company is expecting to gain from it. Also detail to a finer level the exact business processes in terms of who does what, what screens they will utilize, and what areas will require close cooperation.
- *Software installation:* The initial software installation is for the purpose of training only, to give the end users an idea of the look and feel of the system.
- *Key user training:* The key user training is something that should take a while and will include the maintenance process training that was envisaged in the training requirements. The end users must be trained to a level where they are able to operate and understand the system without the software consultants being present at the end of the day.
- *Final software installation and configuration:* At this point, the software is installed in its final form with all required configurations completed. System administrators will need to fully understand the configuration process and its effects on the performance of the system — a critical point in the process.
- *Software testing:* Full, functional testing of the software.
- *Data migrations:* Full data migrations of information either from the old system or from manually prepared information. The intention of the data migration step is to begin using the system with at least some information so as not to be running cold.
- *Go-live tests with end users:* The final of the testing phases, designed to determine if any other errors arise once it is in the hands of the end users.

Post-implementation phase

In the post implementation phase, directly after the go-live test of the system, it is important to conduct regular reviews of the system. The end goal is to ensure that all of the goals and processes are being adhered to and that the system is being fully utilized. Many companies do not use 50% of the CMMS they have purchased, which leads to disillusionment with the system and with the company supplying it.

At all phases during the project there must be strong levels of attention to what is going on. It is very easy in projects of this size and duration to lose the momentum, only to pick it up again later in a piecemeal fashion. By that time, the original impetus has faded and the system is not able to achieve the results required.

The final phase is the re-implementation phase. This generally happens one to two years after the original implementation, and involves a full-scale audit of the system by the software company to see what the levels of use are, and also to see if there are additional gains that can be made by adding areas of the system or focusing on training and business process development. This part of the entire process is a good exercise to see where things can be improved; continuous improvement was why you bought the system in the first place.

A case study

A world-class CMMS was purchased by a top-level petroleum company and implemented using the methodologies and processes discussed in this book (albeit in a less- structured format). After a period of approximately three years, there was a review of the processes to examine the benefits of the implementation and of the system's use. The benefits realized were varied and applied mainly to the areas of logistics and maintenance management.

Maintenance

- High level of maintenance and stores integration due to system
- Increase in useful life spans of equipment due to advanced maintenance methodologies
- Inventory reduction through the use of work order templates and other automation and demand management tools ($3 million USD)
- Higher, enterprise-wide realization of maintenance costs and effects on profitability
- Direct reductions in maintenance costs
- Direct increases in availability
- 40% reduction in corrective maintenance actions
- Increase in mean time between failures and decreases in mean time to repair
- Reduction of maintenance overtime

Inventory management

- Rationalization of inventory levels over multiple sites (reduction of 10,000 items)
- Improvement in reordering costs and times
- Efficiency gains due to interfaces with the Internet

- Efficiency gains in purchasing due to advanced methodologies
- Reduction of inventory management costs due to high use of consignment and vendor-managed stock

As can be seen, the efforts made in this implementation project have paid dividends and in this particular case are continuing to pay dividends.

Above all, due to the approach detailed here in the template system, there is the continual and sustained focus on achieving world-recognized levels of strong performance. There is also the focus to sustain that performance through the different stages of the system's life, and the life of the business.

Index

P

R

S